Aquaculture Biotechnology

Aquaculture Biotechnology

– Author –
Naveen Gupta

Wordpen Academics
Publishing Science That Matter

ISBN: 978-93-90438-13-6 (Hardback)

Published by : **WORDPEN ACADEMICS**
4736/23, Ansari Road, Darya Ganj,
New Delhi – 110 002
Phone: +91-011-35004336
E-mail: info@wordpenacademics.com

Printed at **:** **Replika Press Pvt. Ltd.**

Exclusive Distributor: ***Bio-Green Books, New Delhi***

Preface

This book provides key reviews that look at the application of genetic, cellular, and molecular technologies to enable fish farmers to produce a more abundant, resilient, and healthier supply of seafood. The aquaculture industry has recognized that there can be many benefits gained by incorporating biotechnology and genomics tools and information into their practices. By integrating aquaculture, microbial and engineering expertise, scientists are developing the next generation of marine aquaculture technology that is recirculating, land based, contained and fully compatible with the marine and coastal environments.

Biotechnology - the use of biological systems or living organisms in production processes - has a wide range of useful applications in fisheries and aquaculture. Modern biotechnology is already making important contributions and poses significant challenges to aquaculture and fisheries development. It perceives that modern biotechnologies should be used as adjuncts to and not as substitutes for conventional technologies in solving problems, and that their application should be need-driven rather than technology-driven. The use of modern biotechnology to enhance production of aquatic species holds great potential not only to meet demand but also to improve aquaculture. Genetic modification and biotechnology also holds tremendous potential to improve the quality and quantity of fish reared in aquaculture. It brings opportunities, for instance, to increase growth rate in farmed species, boost the nutritional value of aquafeeds, improve fish health, help restore and protect environments, extend the range of aquatic species and, improve the management and conservation of wild stocks.

The field of genetic biotechnology similarly ranges from simple techniques such as hybridization, to more complex processes such as the transfer of specific genes between species to create GMOs (Genetically Modified Organisms). Over the years our knowledge of fish breeding requirements has improved and the ability developed to induce breeding artificially, through the use of natural or synthetic hormones and/or environmental manipulations. For example changing photoperiod or water temperature can induce some fish to spawn.These have been key factors that have facilitated the application of more advanced biotechnologies. Selective breeding, the maintenance of stocks genetically improved by chromosome manipulation, line crossing, and sex reversal all depend on the controlled breeding of farmed species. The advanced topics in this book offer the readers the opportunity to engage with a wider range of activities within the aquaculture sector.

This book covers all the aspects of this subject. It is hoped, the book will be found to be of immense value to the students of this subject.

—Naveen Gupta

Contents

	Preface	*v*
1.	Biotechnology	1
2.	Biotechnology in Fish Breeding	28
3.	Fish Production: Structure and Composition	45
4.	Necropsy and Diagnostic Techniques	70
5.	Biological Limitations on Transgenic Growth	127
6.	Selection for Quantitative Phenotypes	185
7.	Artificial Hybridization	227
8.	Vertebrate-Derived Compounds	247
	Bibliography	293
	Index	295

Chapter 1

Biotechnology

Biotechnology - the use of biological systems or living organisms in production processes- has a wide range of useful applications in fisheries and aquaculture. It brings opportunities, for instance, to increase growth rate in farmed species, boost the nutritional value of aquafeeds, improve fish health, help restore and protect environments, extend the range of aquatic species and, improve the management and conservation of wild stocks. Some biotechnologies are simple with a long history of application; e.g. fertilization of ponds to increase feed availability. Others are more advanced and take advantage of increasing knowledge of molecular biology and genetics, e.g. genetic engineering and DNA disease diagnosis.

The field of genetic biotechnology similarly ranges from simple techniques such as hybridization, to more complex processes such as the transfer of specific genes between species to create GMOs (Genetically Modified Organisms). Over the years our knowledge of fish breeding requirements has improved and the ability developed to induce breeding artificially, through the use of natural or synthetic hormones and/or environmental manipulations. (For example changing photoperiod or water temperature can induce some fish to spawn).

These have been key factors that have facilitated the application of more advanced biotechnologies. Selective breeding, the maintenance of stocks genetically improved by chromosome manipulation, line crossing, and sex reversal all depend on the controlled breeding of farmed species. These improvements in reproductive technologies have also assisted aquaculturists greatly in their efforts to domesticate aquatic animals. In addition, by making it possible to remove the natural constraints and timing of breeding, farmers are able to mate many more species at times that are most beneficial, and thus help to ensure a steady and consistent supply of fish to the market.

Genetic Biotechnologies

Genetic biotechnologies in aquaculture focus primarily on increasing growth rate. They can also focus on improving disease resistance and increasing environmental tolerance. There are several biotechnologies that can be readily applied to farmed aquatic species.

Selective Breeding

Selective breeding, i.e. traditional animal breeding, started with the common carp several thousand years ago. However, it has only recently been applied to a handful of other species of food fish such as catfish, trout and tilapia. As a result, many farmed aquatic species are still very similar to their wild relatives. Selective breeding programmes have however yielded significant and consistent gains; for instance a 5-20% improvement in growth rate per generation in farmed species such as Atlantic salmon, catfish and tilapia.

Manipulation

Chromosome-set manipulation is a technique that can be used to produce so-called 'triploid' organisms that have three sets of chromosomes instead of the usual two. Triploids generally can not reproduce and so it was initially thought that the energy that was not channelled into reproduction would go instead to increasing growth rate, but this has not in fact proven to be the case. The advantage of triploids seems to be rather in their functional sterility (although that does not reach 100%). For example, triploid oysters do not produce gonads and are therefore marketable at times of the year when mature oysters have an off-taste because of gamete production ('gametes' are sex cells — the ovum, or egg in the female and the sperm in the male).

Monosex Culture

In aquaculture, one sex is often more desirable for the market than the other. For example, female sturgeon produce caviar, male tilapia grow faster than females, female trout and salmon generally grow faster than males. The production of single sex groups of fish takes advantage of this sexual dimorphism and can be accomplished by manipulation of the developing gametes and embryo.

The manipulation can be in the form of DNA denaturisation in the gametes followed by chromosome-set manipulation or alternatively, by hormonal sex-reversal and subsequent breeding.

Administering appropriate hormones can change the phenotypic (physical) sex of many aquatic species. For example, genetically male tilapia can be turned into 'physical' females through estrogen treatments.

These genetic males, that are physically female, are then mated with normal males to produce a group of all-male tilapia that grow faster and have less unwanted matings than a group of mixed-sex tilapia. (In mixed sex culture of tilapia, mating at a small size often leads to overcrowding and stunting).

Some of the all-male offspring of this process have two male chromosomes and these fish can be used in turn as broodstock for subsequent generations. A big advantage of this technique is that all-male populations can be produced in those subsequent generations without the use of any hormones.

Hybridization

Hybridization is another simple genetic technology that has become easier with the development of artificial breeding techniques, such as the use of pituitary gland extract and other hormones to initiate gamete development and induce spawning (the release of fish eggs).

An increased understanding of environmental cues that influence reproduction, such as day length, temperature or water current has also played an important part in improving breeding programmes. Fish farmers can now overcome many of the natural reproductive isolating mechanisms that species develop in the wild. Hybridization can also be used to produce single sex groups of fish when the sex-determining mechanisms in the parental lines are different (for example, hybridization of Nile tilapia and the blue tilapia).

Technological Improvements

These improvements in reproductive technologies have also assisted aquaculturists greatly in their efforts to domesticate aquatic species. In addition, by making it possible to remove the natural constraints and timing of breeding, farmers are able to mate many more species at the times that are most beneficial, and thus help to ensure a steady and consistent supply of fish to the market.

Growth of Transgenic Technology

Genetic engineering is a vague term that has come to be nearly synonymous with gene transfers i.e. the production of transgenic fish or genetically modified organisms (GMOs). This technology is progressing rapidly and it is now possible to move genes between distantly related species; for instance, a gene that produces an anti-freeze protein has been transferred from the winter flounder to strawberries. Gene transfer in fish has usually involved genes that produce growth hormone and has been shown to increase growth rate dramatically in carp, catfish, salmon, tilapia, mudloach and trout. The same anti-freeze gene that was put into plants was put into salmon in the hopes of extending the farming range of the fish. The gene did not produce enough of the protein to extend the salmon's range into colder waters, but it did allow the salmon to continue growing during cold months when non-transgenic salmon would not grow. Transgenic technology is currently in the research and development stage; to our knowledge there are no transgenic aquatic plants or animals available to the consumer.

Genetic Biotechnologies and Health

Genetic biotechnologies are being used to improve fish health through conventional selection for disease resistance and through the use of molecular investigation of pathogens for characterisation and diagnosis. DNA-based technologies are being used now to characterize different species and strains of pathogens. Genetic characterisation of the pathogen may also reveal information about its origin, e.g. DNA analysis revealed two strains of crayfish plague fungus in Sweden: one from the local species and one originating in Turkey. Once the pathogen is characterized, DNA probes can be developed to screen for specific pathogens in tissue, whole animals and even in water and soil samples. These techniques are being used to detect viral diseases of marine shrimp throughout the world and for bacterial and fungal pathogens in fishes in many areas.

Vaccines

Modern biotechnology is also of great value in the field of vaccines and immunostimulants for aquaculture species. These allow preventative measures to be taken to combat disease through vaccination or immunity enhancement. Both can be administered via additives in feeds, immersion or, in the case of the larger culture animals like fish, by injection. Genetically engineered vaccines are also being developed to protect fish against pathogens. Genetic immunisation of rainbow trout with a glycoprotein gene from the virus causing viral haemorrhagic septicaemia has recently been shown to induce high levels of protection against the virus. Work is also underway on immunising carp, salmon and other fishes with genetically engineered vaccines for other diseases. Currently, vaccines are of questionable effectiveness in crustacea and may be difficult or too costly to use in developing countries.

Detecting Human Pathogens

The type of DNA 'probes' that are used to screen for pathogens that affect fish or shrimp could also be developed to check for pathogens that affect humans. This will become increasingly important in answering concerns in the market place about the safety of aquatic products for human consumption. These new molecular techniques are extremely sensitive and can identify pathogens in fish long before there are any clinical signs of the disease. This has implications for quarantine and the trade of aquatic species, which is currently governed by the World Trade Organisation and the Office International des Epizooties. Trade can be restricted based on the disease status of a product or a region; identification of minute quantities of a pathogen or of a new strain of an existing pathogen could change or influence existing trade patterns.

Bio-Remediation

Farmed aquatic animals are much more sensitive to their immediate environment than land animals. The water they are immersed in, and on which they depend for oxygen and a range of other important chemicals, also takes up their waste products and may carry pollution from the nearby environment. The process of disease in aquaculture species is thus much more strongly connected to environmental factors than would be the case say, with cattle. A further biotechnology field that has developed in aquaculture, because of the nature of this relationship, is that of bio-remediation. This refers to the use of 'friendly bacteria or 'pro-biotics' to treat water or feeds and, by natural processes, discourages the development of 'unfriendly' bacteria that potentially would cause disease. Many such products are currently being marketed' — while conclusive studies have yet to be carried out to check their effectiveness, some of them do seem to bring production benefits.

Genetically Modified fish

Genetically modified fish (GM fish) are genetically modified organisms. The DNA of the fish has been modified using genetic engineering techniques. In most cases the aim is to introduce a new trait to the fish which does not occur naturally in the species. GM fish are used in scientific research, and while they are being developed for use in aquaculture food production, as of May 2012 no GM fish has been approved by the FDA for this purpose. Some GM fish that have been created have promoters driving an over-production of "all fish" growth hormone. This resulted in dramatic growth enhancement in several species, including salmonids, carps andtilapias.

Critics have objected to GM fish per se on several grounds, including ecological concerns, and with respect to whether using them as food is safe and whether GM fish are needed to address the world's food needs. See the genetically modified food controversies article for discussion of issues about GM food and Regulation of the release of genetic modified organisms for discussion of regulatory regimes around the world.

History and Process

The history and key methods of genetic engineering are discussed on that page in detail. Generally, genetic modification entails manipulation of DNA. The process is known as cisgenesis when a gene is transferred between organisms that could be conventionally bred, or transgenesis when a gene from one species is added to a different species. Gene transfer into the genome of the desired organism, as for fish in this case, requires a vector like a lentivirus or mechanical/physical insertion of the altered genes into the nucleus of the host by means of a micro syringe or a gene gun.

The first recorded instances of production of transgenics in aquatic species are those of Maclean and Talwar (1984) in rainbow trout and Zhu et al. (1985) in goldfish. Since then over 35 species have been genetically engineered in research laboratories.

Types

Salmon

Salmon belong to the Salmonidae family which also includes salmon and trout. Although the smallest species is just 13 centimetres (5.1 in) long as an adult, most are much larger, and the largest can reach 2 metres (6.6 ft). All salmonids spawn in fresh water, but they spend most of their maturity in the sea. This life style is known as anadromous. They are considered to be predators, because they feed on small crustaceans, aquatic insects, and smaller fish. A genetically modified Atlantic salmon known as the AquAdvantage salmon has an increased growth rate and size over the wild type Atlantic salmon from which it was derived, up to doubling its weight with a reduced time of growth to maturity. Although materials have been submitted to obtain approval to grow and market the AquaAdvantage salmon, as of December 2012 the FDA had not granted approval.

Tilapia

Tilapia is the common name for several species of cichlid fish from the tilapine cichlid tribe. Tilapia inhabit a wide range fresh water habitats, including lakes, streams, ponds and rivers. Anciently, tilapia hold great significance inartisan fishing in Africa, and are paramount lately in aquaculture. Tilapia are very vulnerable to cold temperatures, and thus survive well with temperatures above 60 °F (16 °C). (See tilapia as exotic species.) Tilapia is the fifth most important fish in fish farming, with production reaching 1,505,804 metric tons in 2000. Because of their large size, rapid growth, and palatability, tilapiine cichlids are the focus of major farming efforts, specifically various species.

Zebrafish

Zebrafish are freshwater fish and are part of the Cyprinidae. They are a popular aquarium fish, commonly sold as zebra danio, and have been very vital as model organisms in research. They derive their name from the uniform horizontal stripes along the side of the body bilaterally. Males bear gold stripes within the blue stripes, while females bear silver stripes within the blue stripes. Zebra fish can mature up to 6.4 centimetres in the wild, but usually it is rare for them to mature beyond 4" in captivity.

Uses - Actual and Potential

Research

Most genetically engineered fish are used in basic research in genetics and development. Two species of fish, zebrafish and medaka, are most commonly modified because they have optically clear chorions (shells), rapidly develop, and the 1-cell embryo is easy to see and microinject with transgenic DNA. Also, zebrafish have the capability of regenerating their organ tissues, and GM zebrafish are being explored for benefits of unlocking human organ tissue diseases and failure mysteries. For instance zebrafish are used to understand heart tissue repair and regeneration in efforts to study and discover cures for cardiovascular diseases.

Pets

The GloFish is a patented brand of genetically modified (GM) fluorescent zebrafish with bright red, green, and orange fluorescent colour. Although not originally developed for the ornamental fish trade, it became the first genetically modified animal to become publicly available as a pet when it was introduced for sale in 2003. They were quickly banned for sale in California.

Food (Potential)

Genetically modified fish have been developed with promoters driving an over-production of "all fish" growth hormone for use in theaquaculture industry to increase the speed of development and potentially reduce fishing pressure on wild stocks. This has resulted in dramatic growth enhancement in several species, including salmon, trout and tilapia.

AquaBounty, the leading company in GM fish for the food industry, claims that their GM AquAdvantage salmon can mature in half the time it takes non-GM salmon and achieves twice the size. AquaBounty has applied for regulatory approval to market their GM salmon in the US. As of May 2012 the application was still pending.

Detecting Aquatic Pollution (Potential)

Several academic groups have been developing GM zebrafish to detect aquatic pollution. The lab that originated the GloFish discussed above originally developed them to change colour in the presence of pollutants, to be used as environmental sensors. A lab at University of Cincinnati has been developing GM zebrafish for the same purpose, as has a lab at Tulane University.

Regulation

The regulation of genetic engineering concerns the approaches taken by governments to assess and manage the risks associated with the development and release of genetically modified crops. There are differences in the regulation of GMOs between countries, with some of the most marked differences occurring between the USA and Europe. Regulation varies in a given country depending on the intended use of the products of the genetic engineering. For example, a fish not intended for food use is generally not reviewed by authorities responsible for food safety.

Controversy

Critics have objected to use of genetic engineering per se on several grounds, including ethical concerns, ecological concerns (especially about gene flow, and economic concerns raised by the fact GM techniques and GM organisms are subject to intellectual property law. GMOs also are involved in controversies over GM food with respect to whether using GM fish as safe is safe, whether it would exacerbate or cause fish allergies, whether it should be labelled, and whether GM fish and crops are needed to address the world's food needs. These controversies have led to litigation, international trade disputes, and protests, and to restrictive regulation of commercial products in most countries.

With respect to concerns about GM AquAdvantage Salmon interbreeding with wild fish, the company indicates that their GM salmon are not capable of reproducing, as the GM fish are sterile. AquaBounty also emphasizes that their GM fish would not survive wild conditions due to the geographical locations where their research is being done, as well as the locations of their farms.

An article in Slate Magazine in December 2012 by Jon Entine, Director of the Genetic Literacy Project, criticized the Obama Administration for preventing the publication of the environmental assessment (EA) of the AquAdvantage Salmon, which was completed in April 2012 and which concluded that "the salmon is safe to eat and poses no serious environmental hazards."

The Slate article said that the publication of the report was stopped "after meetings with the White House, which was debating the political implications of approving the GM salmon, a move likely to infuriate a portion of its base". Within days of the article's publication and less than two months after the election, the FDA released the draft EA and opened the comment period.

Gmos in Aquatic Species

The first transgenic animal to be produced was a mouse (Palmiter, Brinster and Hammer, 1982). The first recorded instances of production of transgenics in aquatic species are those of Maclean and Talwar (1984) in rainbow trout and Zhu *et al.* (1985) in goldfish. Since then many species have been used to produce GMOs as shown in Table 1. The list represents an amalgam of species significant in aquaculture with species amenable to laboratory culture and with short life cycles used particularly for studies of gene action, studies which of course form the platform for better understanding and hence better production in aquaculture.

Table: *Aquatic species in which GMOs have been induced*

Common name	***Latin name***	***Number of constructs employed to generate transgenics***
Fish		
Atlantic salmon	*Salmo salar*	6
Coho salmon	*Oncorhynchus kisutch*	4
Chinook salmon	*Oncorhynchus tshawytscha*	3
Tilapia	*Oreochromis spp.*	12
Medaka	*Oryzias latipes*	17
Zebra fish	*Brachydanio rerio*	14
Common carp	*Cyprinus carpio*	14
Channel catfish	*Ictalurus punctatus*	9
African catfish	*Clarias gariepinus*	1
Rainbow trout	*Oncorhyncus mykiss*	7
Cutthroat trout	*Oncorhyncus clarki*	1
Goldfish	*Carrassius auratus*	5
Northern pike	*Esox lucius*	2
Loach	*Misgurnus anguillicaudatus*	2
Sea bream	*Sparus aurata*	2
Red Sea Bream	*Pagrus major*	1
Blunt snout bream	*Megalobrama amblycephala*	1
Nigorobuna	*Carrassius auratus grandoculis*	1
Walleye	*Stizostedion vitreum*	1
Others		
Brine shrimp	*Artemia franciscana*	1
Seaweed	*Laminaria japonica*	1
	Undaria pinnatifida	
Sea Urchin	*Strongylocentrotus purpuratus*	1
	Paracentrotus lividus	
	Arbacia lixula	
Abalone	*Haliotus rufescens*	1

It is clear that Atlantic and coho salmon, tilapia species, catfish, medaka and zebrafish dominate in terms of numbers. Of these fish groups three are very important in aquaculture.

The Process of Genetic Modification

Production of GMOs is a multistage process which can be summarized as follows:

1. identification of the gene interest;
2. isolation of the gene of interest;
3. amplifying the gene to produce many copies;
4. associating the gene with an appropriate promoter and poly A sequence and insertion into plasmids;
5. multiplying the plasmid in bacteria and recovering the cloned construct for injection;
6. transference of the construct into the recipient tissue, usually fertilized eggs;
7. integration of gene into recipient genome;
8. expression of gene in recipient genome; and
9. inheritance of gene through further generations.

Isolation of the Gene of Interest

Usually the gene of interest will already be available as an element of a "library" of short sections of the total genome of the donor strain or species. If this is the case the procedure followed is to multiply the gene using the PCR reaction. If, however, the gene is to be taken from a genome not previously investigated, a more complex procedure will need to be followed. The use of the technique of the Polymerase Chain Reaction (PCR) enables the gene in both the cases noted above to be multiplied to the level of several million copies needed for the generation of the construct.

Cloning the Gene of Interest

When many copies of the target gene have been generated, the gene is placed in a "construct". Once the gene of interest has been ligated enzymatically into the construct, this whole complex is ligated into bacterial plasmids, which act as "production vectors" and enable the gene to be replicated many times within the bacterial cells. The bacteria are then plated out. It is possible to tell from reporter genes whether the vector has been taken up by the bacterial cells. This usually involves some colour change in the colonies containing inserted DNA. The many times amplified DNA construct is then enzymatically cut out of the plasmids (after these have been removed from the bacterial cells) and it is ready to be used for insertion into eggs of the host species.

A more detailed outline of the technical details of the processes outlined in Sections 5.2 and 5.3 may be found in Maclean (1998).

The Construct

A construct is a piece of DNA which functions as the vehicle or vector carrying the target gene into the recipient organism. It has several different regions as shown in Figure. There is a promoter region which controls the activity of the target gene, a region where the target DNA is inserted, usually some type of reporter gene to enable one to ascertain whether the target has combined successfully with the construct and a termination sequence.

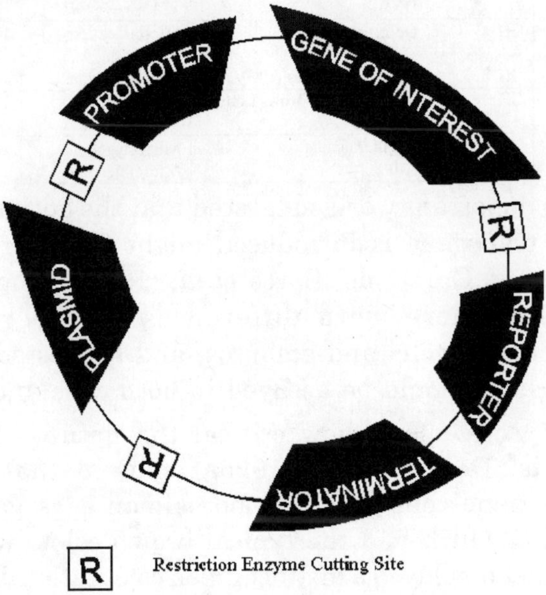

Figure: *Diagram of DNA sequence of a basic plasmid and incorporated construct.*

The sources of these several DNA sequences may be different species although promoter and target genes would ideally be derived from the same species

As shown in Table 3, constructs have been reported from 92 studies. The number of different constructs is greater than the number of target genes used in aquaculture and a substantial research effort has been made in this area. From the early 1990s research focussed on developing "all fish" constructs in preference to using mammalian promoters.

The use of all-fish constructs has dramatic effects on expression of transgenes, e.g. Devlin *et al*. (1994), developed an all salmon gene construct which accelerates the growth of transgenic salmonids by over 11 fold. In tilapia, Maclean (1994) found that using carp beta actin instead of rat beta actin promoter led to a ten fold increase in production of hormone in transgenic animals.

Table: *Summary of major research effort in inducing GMOs in aquatic species.*

Species	*Target gene*	*Typical construct*	*Typical induction method*	*Number of studies*
Salmon spp.	GH AFP	Ocean pout AFP linearized DNA	Microinjection	17/92
Rainbow Trout	GH	Ocean pout AFP	Microinjection	14/92
Tilapia spp.	GH	Cytomegalovirus (CMV)	Microinjection	12/92
Carp	GH	Rous Sarcoma Virus Long Tandem Repeat	Microinjection	17/92
Zebrafish	Luciferase	pMTL plasmid	Microinjection	16/92
Medaka	CAT	AFP	Microinjection	11/92

Other important work suggested that the optimal stage at which the transgene is introduced might vary between cells and species eg. Garcia del Barco *et al.* (1994) using Zebrafish showed that there were differences in the regulatory requirements for cells and embryos, and suggested therefore that constructs should be assayed in both cells and embryos.

Other work shows how critical the nature of the gene construct is. Devlin *et al.* (1995a) showed that using an opAFPGHc gene construct in coho salmon eggs gave rise to some alevins which had the typical brown colouration, while the remainder displayed a distinct green colouration. The results suggest that the green phenotype arose from the presence of the opAFPGHc construct and therefore could be indicative of transgene uptake/transmission. All the offspring were tested by PCR for presence of the transgene and 182 of 184 alevins were correctly assigned on this basis.

However, it was found that later in development all fish turned green (the normal colour later in development) and so the transgenic fish were showing accelerated growth. Later in development it was found that most of the transgenic fish showed signs of cranial abnormality probably due to accelerated growth. While the onstruct was useful in that transgene uptake could be monitored, further work was needed to ensure that healthy fish could be produced.

Techniques for Inducing Transgenics

Transgenic fish have largely been produced through microinjection into fertilised eggs or early embryos. Electroporation of sperm has been shown to be successful in some species eg. Zebrafish (Khoo *et al.,* 1992) Chinook salmon (Sin *et al.*, 1994) and Loach (Tsai, Tseng and Liao, 1995). Liposomes have also been utilised as vectors (Khoo 1995). Ballistic methods using microprojectiles have been investigated in *Artemia* with a view to their use in generating transgenic crustacea (Gendreau *et al.*, 1995) and also in seaweed species (Qin *et al.*,1994). "Baekonisation", an electric, flat field type of electroporation was utilised to transfer DNA into Zebrafish embryos (Zhao, Zhang and Wong, 1993), this method appeared to be successful but has not been taken up in the same way as other forms of electroporation and microinjection methods.

More recently the use of embryonic stem cells (ESC) as a method for inducing transgenesis has been advocated. These cells are undifferentiated and remain totipotent, so they can be manipulated in vitro and subsequently reintroduced into early embryos where they can contribute to the germ line of the host. In this way genes could be stably introduced or deleted (Melamed *et al.*, 2002). Despite the early success of ESC technology in mice, the uptake of the technology for fish has been slow, although early precursor cells (Mes 1) have been cultivated from Medaka and show many of the same features as mouse ESC. Studies by Hong, Winkler and Schartl (1996, 1998) and Hong, Chen and Schartl (2000) showed that 90 percent of host cell blastulae transplanted with Mes 1 cells developed into mosaic fry, and these cells became integrated into organs derived from all three germ layers, and differentiated into various types of functional cells.

Another example of new and possibly more efficient ways for gene transfer is the use of pantropic retroviral vectors. These are able to infect a wide range of host cells and have been used to infect newly fertilized Medaka eggs with a reporter gene, which appeared to become integrated into the entire germ line of some of the P1 females (Lu, Burns and Chen, 1997). In Zebrafish when retroviral infection and microinjection were compared, the two methods were equally efficient in passing the transgene into eggs, but there was wider variability in the extent of reporter gene expression among those founders that were microinjected (Linney *et al.*,1999). However, the use of retroviruses is not without problems.

The microinjection method is suitable for relatively small numbers of organisms being manipulated whereas electroporation, sperm/liposome mediation and bombardment methods are more suitable for mass treatments. The most popular method of insertion of transgenes in aquaculture is microinjection; in 92 studies reviewed from 1985 to the present, 68 used microinjection, eleven used sperm mediated methods, six used electroporation and five used both sperm mediation and electroporation. However, the problem of mosaic expression of the transgenes is common, and this gives rise to varying proportions of transgenic genotypes in the progeny.

A useful review of technical details of the techniques mentioned can be found in Sin (1997).

Integration Sites

The factors determining sites of integration are still poorly understood though research in this direction is increasing. It is particularly important to gain greater accuracy in controlled site of integration because of the unpredictable effects of uncontrolled integration on resident genes. Caldovic and Hackett (1995) tested the ability of special sequences called transposable border elements from other species to confer position-independent expression of transgenes or enhance integration of transgenic constructs into fish chromosomes. Early results indicate that such elements from some species do not act as enhancers and do not improve integration frequencies. However, both avian and insect border elements were found to confer position-independent expression as judged from expression of CAT genes in F_1 fish. Hackett *et al.*, (1994) showed that co-transfer of retroviral integrase protein with transgenic DNA can accelerate and enhance the rate of integration. More studies of this type are needed to improve the success and controlled positioning of integration of transgenes in the future.

Expression of Gene

The uptake and integration of a transgene does not guarantee that the gene will express itself in the new genetic environment. Tests must be carried out to determine whether there is expression and if there is expression, at what level this takes place. Clearly, in commercial aquaculture only those transgencs expressing the target gene at a sufficiently high level will be of interest.

Inheritance of Gene

A fish which expresses the target gene at an acceptable level may not be able to transmit the gene to progeny. This is because many transgenics are mosaic individuals and unless the gonads are included in the tissues possessing the transgene the transgenic animals will not breed true. Appropriate breeding tests must, therefore, be carried out.

The high proportion of mosaic individuals is one reason why the proportions of progenies of different genotypes resulting from parents that are putatively hemizygous for a transgene do not necessarily conform to mendelian expectations. Another reason is the integration of two or more copies of the transgene at different sites in the recipient genome. Further breeding tests will be required in order to establish a pure breeding line of transgenic fish.

Risk Factors of Gmos

There are a number of publications which address this issue. Maclean and Laight (2001) and Dunham (1999) have produced very useful reviews which discuss many of the points raised in this paper.

In our view the most important areas of risks which need to be considered in the use of transgenics are:

1. human health
2. biodiversity
3. animal welfare
4. poor communities

In each of these categories there exists a multiplicity of pathways by which effects could, in principle, be brought about. Rational and responsible assessment of risk requires that the following properties are all considered:

1. source of the DNA of the target gene;
2. source of the non target DNA segments of the construct used;
3. site(s) of incorporation of the transgene within the recipient genome;
4. product of the transgene;
5. interaction of the transgenic product with other molecules in host and consumer;
6. possible molecular changes in transgene product during processing;
7. pleiotropic effects of transgene;
8. tissue specificity of transgenic expression; and
9. numbers of transgenic organisms capable of interacting with natural systems).

Human Health

The risks to health will depend upon all of the factors listed above. In practical terms the most important of these are likely to be the source of the DNA and the nature of the product. The great majority (98 percent) of dietary DNA is degraded by digestive enzymes relatively quickly (Royal Society, 2001) but use of viruses (disarmed or otherwise) as vectors, must increase the risk factor significantly as these are organisms which are adapted to integrating into host genomes and some represent risk factors for cancer induction. The work of Zhixong Li *et al.* (2002) who induced leukaemia by using retroviral vectors in making transgenics for a commonly used marker gene in mice and a recent report of leukaemia induction in a child undergoing gene therapy for x-SCID using a retrovirus (Hawkes, 2002) show that this is not a trivial risk. Arguments about risks and benefits attached to this form of gene therapy are current (Kaiser, 2003).

At the other extreme the use of autotransgenics must be seen as posing a risk which is orders of magnitude lower than that for allotransgenics and probably negligible. The major risk from the production of the transgene will lie in the use of novel proteins or other molecules produced by the transgenic organisms. Either in the native form or, following modifications in the human body, such molecules could be inimical to human health (e.g. through allergies). It would seem sensible to avoid the use of such substances except where strictly necessary and under rigorous control.

Other potential risks may lie in incorporation of transgenic DNA into the genomes of resident gut microflora (though this is likely to be very improbable) or a change in the pathogen spectrum of the transgenic fish leading to it hosting a new pathogen which happens to be also a human pathogen. Maclean and Laight (2000) assessed risks to consumers as "very low".

Biodiversity

The extent of aquatic diversity is both extremely large and relatively poorly understood (Beardmore, Mair and Lewis, 1997). This means that the task of estimating the risks to aquatic biodiversity at all of its levels from the use of GMOs or indeed, any genetically distinctive strain used in aquaculture is monumentally large.

Aquaculture has a further problem in that the (almost always unintended) escapes of genetically distinct farmed fish are unpredictable and often large in numbers. Stenquist (1996) in discussing transgenics in open ocean aquaculture, quotes some relevant figures. Thus, 15 percent escapes for Atlantic salmon, escapes of 150 000 salmon and 50 000 trout in Chile and catch statistics for Atlantic salmon off Norway in which 15?20 percent of the fish caught were of farmed origin. In Scotland an escape of 100 000 Atlantic salmon was reported recently. It is clear that escapes of these magnitudes pose considerable problems and it is not surprising that in some parts of Norway fish of farmed origin represent a majority of the animals fished (Saegrov *et al.*, 1997)

The major focus of attention in the literature lies, understandably, upon the effects of escapes upon natural populations of the same species, but we must always bear in mind possible impacts across an assemblage or ecosystem as a whole. The first general point to make is that there is, in principle, no difference between the biodiversity risks from escapes of GMOs and from fish genetically improved in some other way, e.g. by selective breeding or (in some respects) from exotic species.

The second general principle is that such genetically improved forms including GMOs, are developed for a specific set of environmental circumstances in which they enjoy an advantage conferred by human decisions. In nature, however, such genetically distinct forms may legitimately be regarded as mutant forms of the wild type. A considerable body of genetical knowledge tells us that the probability of survival of mutant forms is extremely low because they are disadvantaged in viability and/or fertility under natural conditions. Thus, for example, in the genetically distinct farmed Atlantic salmon in Norway the males are very much less successful than wild males in securing mates (Jonssen, 1997).

However, it must be conceded that in species like salmon where the farmed populations outnumber the wild populations by orders of magnitude, the effects of escapes of any genetically distinct genotype upon natural populations may be both deleterious and of significant size simply as a result of "swamping"

An interesting model of the effects on a medaka (*Oryzias latipes*) population of transgenic release has been produced by Muir and Howard (2001) using estimates of juvenile and adult viability, age at sexual maturity, female fecundity, male fertility and mating advantage. They were able to demonstrate that the transgene would spread in natural populations, despite low juvenile viability, if transgenes have sufficient high positive effects on other fitness components. It has been argued that this might lead to extinction but the selective pressure for recombinant genomes with higher viability would be expected to be immense. Maclean and Laight (2000) simulated the changes in frequency of a transgene expected with different scenarios embracing a range of selective values including heterozyote advantage. They note that "repeated small introductions [of the transgene] can have an effect on ... frequency ... since the frequency of advantageous alleles rises much more rapidly than if a single large introduction is considered".

A major problem in assessing risk to natural populations is that of scale. Even if farmed fish are at a selective disadvantage in natural conditions, the ratio of wild:farmed numbers may in some areas, be relatively small. In these situations significant modification of the "native" population and its role in the ecosystem is inevitable.

Whilst not providing a completely satisfactory answer, there is little doubt that making farmed fish sterile would go a long way towards reducing the pressure upon such threatened ecosystems. A number of research efforts to develop systems for sterile fish production are being made. The techniques include triploidisation, antisense transgenics, ribozymes and gene targeting (Maclean, 2002; Uzbekova *et al.*, 2001; Maclean, pers. com.).

Provided that the best containment measures (physical and biological) are adopted, in our opinion, in general risks to biodiversity by GMOs *per se* are probably extremely small, but in specific cases, the risks and consequences may be large. As a general rule and adopting a precautionary approvah (OECD, 1995), it is, however, clear that each individual case needs careful study and appraisal and the best possible containment measures before approval for uptake into commercial production is given.

Animal Welfare

The direct or indirect effects of transgenesis upon the welfare of fish GMOs in aquaculture are very poorly understood. In part, no doubt, this is because notions of cruel or unnatural treatment in mammalian species translate, for a variety of reasons, imperfectly to fish. Nevertheless, as life forms with highly developed nervous systems and with a range of behavioural phenotypes which flow from this, fish qualify for welfare consideration. There are a few studies which bear on this. Thus, for example, Devlin *et al.* (1995b) reported changes in colouration, cranial deformities and opercular overgrowth and lower jaw deformation in coho salmon transgenic for AFP and GH. After one year of development anatomical changes due to growth of cartilage in the cranial and opercular regions were more severe and reduced viability was evident.

The larger body of data on species farmed terrestrially shows dysfunctional development leading to acromegaly, lameness and infertility in some GH transgenics in pigs and sheep. However, in pigs dietary modification influencing nutritional levels of zinc proved successful in avoiding such abnormalities (Pursel and Solomon, 1993; Pursel, 1998).

We have been unable to find systematic data on the incidence, in fish GMOs, of effects such as those described by Devlin *et al.* (1995b) and this is probably because animal welfare is not sufficiently widely recognised as an issue in relation to the use of GMOs.

This is well illustrated in the otherwise comprehensive and balanced review by Sin (1997) in which the section on ethical issues contains no reference to animal welfare. Nevertheless, if GMOs are to be used in aquaculture (and there are weighty arguments for so doing), concerns on this issue will need to be properly satisfied. The Royal Society report (2001) devotes a significant amount of space to this issue.

Poor Communities

This term rather than poor countries is used because all poor countries contain rich people and rich communities. The possible economic disadvantages of use of transgenics centre on two issues:

Dependence on External Agencies for Seed Fish

If transgenic fish become widely grown because they are much more efficient, and if special broodstock are required to produce fry for on-growing to adults, which, cannot be used as broodstock, a dependency is created. This dependency may be benign or oppressive, depending on the arrangements made for seed supply.

Intellectual Property Rights

This is a very difficult issue indeed. Since genes may now be patented and therefore, enjoy commercial value, the opportunities for dispute about equitable treatment of stakeholders in cases where ownership of genes and strains is contested, are legion.

A recently published report (Commission on Intellectual Property Rights, 2002) states that developing countries are frequently disadvantaged in the use of, and access to, IPR because of increasingly protective attitudes taken by owners of IPR. However, the report also indicates that developing countries are very heterogeneous in respect of their ability to use and develop IPR.

Regulation, Policy and The Climate of Acceptance

The extent to which GMOs are perceived as desirable for aquaculture and food supply has been probed by Bartley and Hallerman (1995) on a questionnaire basis. The responses were generally positive in terms of exploring the potential of transgenesis. It is clear however, that in human populations as a whole there is a severe deficiency of knowledge appropriate for making informed decisions about the value of GMOs (Dunham, 1999).

The task of instituting and managing well thought out, responsible and scientifically sound measures is made more difficult by the frequently irresponsible and inaccurate media treatment of GMOs on the one hand and the cavalier pronouncements by some authorities on the other hand, e.g. that no distinction need by made in labelling between food derived from GMOs and from non-GMOs. Such extreme differences in attitude tend to inflame public opinion.

The need for incorporation of risk assessment and risk management procedures relating to use of GMOs in aquatic species has been well brought out by Hallerman and Kapuscinski (1995). According to Hallerman and Kapuscinski (1995), "as a generality among developed countries at least, the public will support biotechnology if it yields a healthful product in an environmentally sound manner. This statement encapsulates implicitly a most significant factor - that in considering the benefits and disadvantages of GMOs, perception by the general public is a most important factor in shaping attitude of regulatory agencies" (emphasis by JAB and JP).

There is at least a suspicion that some contemporary science may also be affected by perceptions. A recent furore over the alleged presence, in traditional varieties of maize in Mexico, of transgenes derived from modern strains of cultivated maize points to this. The matter is still not completely resolved but has led to *Nature* declaring it should not have published the original paper because of technical inadequacies (Mann, 2002).

It is perhaps dangerous to extrapolate directly from agronomy to aquaculture but it is worth noting that, in the USA, which accounts for more than two thirds of the acreage planted to crop GMOs, moves to institute a much tighter regulatory framework are in train (Soil Association 2002). Five bills were introduced in May 2002 in the House of Representatives in Washington to cover legal protection for farmers, increase GM food safety, require labelling for foods containing, or produced with, GMOs, address developing country issues and assign liability for damages.

These measures appear to have sprung, at least in part, from serious dissatisfaction on the part of some farmers with GM crop plants. It is, however, relevant to note that the dominant position enjoyed by several large multinational agrochemical companies in providing both seed and agrochemicals is not reflected in the aquaculture industry which is far more fragmented. Nevertheless, the indications are already present that there is a move towards fewer and larger companies within the sector and this could have consequences for the control and use of GMOs in aquaculture species.

A fair number of countries have instituted regulatory arrangements for the culture, release and dietary utilisation of GMOs, but with considerable differences in the approaches and restrictions employed (discussed by Maclean, 1998).

At the international level the Cartagena Biosafety Protocol of the Convention on Biological Diversity provides a comprehensive and rigorous framework for regulation of protection of biodiversity and argues that this be done "taking into account risks to human health and specifically focussing on transboundary movements". The great majority of nations are signatories to the convention. What is less clear is the extent to which, at national and regional levels, the protocols set out are implemented and implementable through domestic legislation and effective sanctions.

Among its provisions, the Cartagena protocol has a detailed treatment of risk assessment in relation to the protection of biological diversity. Among the important and relevant general principles are:

1. "Risk assessment should be carried out on a case by case basis."
2. "Risks associated with LMOs should be considered in the context of the risks posed by the non-modified organisms in the receiving environment." and in the [comparable] methodology of risk assessment:
    1. "An estimation of the overall risk posed by the LMO based on an evaluation of the likelihood and consequences of the adverse risks being realised."
    2. "A recommendation as to whether or not the risks are acceptable or manageable, including, where necessary, identification of strategies to manage these risks."

We are, thus, faced with a complex array of stakeholders; breeders and growers, geneticists, multinational companies, supermarkets, consumers, politicians and the media. All have their own agenda to follow, but critically important to all of them and to a reasonable resolution of the current conflict is the availability and use of full, sound and accurate information, particularly in the context of framing of appropriate regulations affecting GMOs

Biotechnology and Its Applications in Aquaculture and Fisheries

Biotechnology provides powerful tools for the sustainable development of aquaculture, fisheries, as well as the food industry. Increased public demand for seafood and decreasing natural marine habitats have encouraged scientists to study ways that biotechnology can increase the production of marine food products, and making aquaculture as a growing field of animal research. Biotechnology allows scientists to identify and combine traits in fish and shellfish to increase productivity and improve quality. Scientists are investigating genes that will increase production of natural fish growth factors as well as the natural defence compounds marine organisms use to fight microbial infections.

Modern biotechnology is already making important contributions and poses significant challenges to aquaculture and fisheries development. It perceives that modern biotechnologies should be used as adjuncts to and not as substitutes for conventional technologies in solving problems, and that their application should be need-driven rather than technology-driven.

The use of modern biotechnology to enhance production of aquatic species holds great potential not only to meet demand but also to improve aquaculture. Genetic modification and biotechnology also holds tremendous potential to improve the quality and quantity of fish reared in aquaculture. There is a growing demand for aquaculture; biotechnology can help to meet this demand.

As with all biotech-enhanced foods, aquaculture will be strictly regulated before approved for market. Biotech aquaculture also offers environmental benefits. When appropriately integrated with other technologies for the production of food, agricultural products and services, biotechnology can be of significant assistance in meeting the needs of an expanding and increasingly urbanized population in the next millennium.

Successful development and application of biotechnology are possible only when a broad research and knowledge base in the biology, variation, breeding, agronomy, physiology, pathology, biochemistry and genetics of the manipulated organism exists. Benefits offered by the new technologies cannot be fulfilled without a continued commitment to basic research. Biotechnological programmes must be fully integrated into a research background and cannot be taken out of context if they are to succeed.

Indian fisheries and aquaculture is an important sector of food production, providing nutritional security to the food basket, contributing to the agricultural exports and engaging about fourteen million people in different activities. With diverse resources ranging from deep seas to lakes in the mountains and more than 10% of the global biodiversity in terms of fish and shellfish species, the country has shown continuous and sustained increments in fish production since independence. Constituting about 4.4% of the global fish production, the sector contributes to 1.1% of the GDP and 4.7% of the agricultural GDP. The total fish production of 6.57 million metric tonnes presently has nearly 55% contribution from the inland sector and nearly the same from culture fisheries.

Fish and fish products have presently emerged as the largest group in agricultural exports of India.(Marine products export review-MPEDA., April 2006-March2007). The potential area of biotechnology in aquaculture include the use of synthetic hormones in induced breeding, transgenic fish ,gene banking , uniparental and polyploidy population and health management.

Chapter 2

Biotechnology in Fish Breeding

Gonadotropin releasing hormone (GnRH) is now the best available biotechnological tool for the induced breeding of fish. GnRH is the key regulator and central initiator of reproductive cascade in all vertebrates (Bhattacharya et al.,2002). It is a decapeptide and was first isolated from pig and ship hypothalami with the ability to induce pituitary release of luteinising hormone (LH) and follicle stimulating hormone (FSH) (Schally et al.,1973). Since then only one form of GnRH has been identified in most placental mammals including human beings as the sole neuropeptide causing the release of LH and FSH. However ,in non mammalian species (except guinea pig) twelve GnRH variants have now been structurally elucidated ,among them seven or eight different forms have been isolated from fish species.(Halder et al.,1991;Sherwood et al.,1993;King and Miller,1995;Jimenez-Linan et al.,1997). The most recent GnRH purified and characterized was by Carolsfeld et al.(2000) and Robinson et al.(2000). Depending on the structural variant and their biological activities, number of chemical analogues have seen prepared and one of them is salmon GnRH analogue profusely used now in fish breeding and marked commercially under the name of ýÿOvaprimýÿ throughout the world .The induced breeding of fish is now successfully achieved by development of GnRH technology.

Transgenesis

Transgenesis or transgenics may be defined as the introduction of exogenous gene / DNA into host genome resulting in its stable maintenance, transmission and expression. The technology offers an excellent opportunity for modifying or improving the genetic traits of commercially important fishers, mollusks and crustaceans for aquaculture. The idea of producting transgenic animals became popular when Palmitter et al. (1982) first produced transgenic mouse by introducing metallothionein human growth hormone fusion gene (mT-hGH) into mouse egg, resulting in dramatic increase in growth. This triggered a series of attemptson gene transfer in economically important animals including fish.

The first transgenic fish was produced Zhu et al. (1985) in China, who claimed the transient expression n putative transgenics, although they gave no molecular evidence for the integration of the transgene. The technique has now seen successfully applied to a number of fish species. Dramatic growth enhancement has been shown using this technique especially in salmonids (Devlin et al., 1994). Some studies have revealed enhancement of growth in adult salmon to an average of 3 ýÿ 5 times the size of non ýÿ transgenic controls, with some individuals, especially during the first few months of growth, reaching as much as 10 ýÿ 30 times the size of the controls (Devlin et al., 1994; Hew et al., 1995).

The introduction of transgenic technique has simultaneously put more emphasis on the need for production of sterile progeny in order to minimize the risk of transgenic stocks mixing in the wild populations. The technical development has expanded the possibilities for producing either sterile fish or those whose reproductive activity can be specifically turned on or off using inducible promoters.

This would clearly be of considerable value allowing both optimal growth and controlled reproduction of the transgenic stocks while ensuring that any escaped fish would be unable to breed. An increased resistance of fish to cold temperatures has been another subject of research in fish transgenics for the past several years (Fletcher et al., 2001). Coldwater temperatures pose a considerable stressor to many fish and few are able to survive water temperatures much below 0-1°C. this is often a major problem in aquaculture in cold climates. Interestingly, some marine teleosts have high levels (10 ýÿ 25 mg/ml) of serum antifreeze proteins (AFP) or glycoproteins

(AFGP) which effectively reduce the freezing temperature by preventing ice-crystal growth. The isolation, characterization and regulation of these antifreeze proteins particularly of the inter flounder Pleuronectas americanus has been the subject of research for a considerable period in Canada. Consequently, the gene encoding the liver AFP from winter flounder was successfully introduced into the genome of Atlantic salmon where it became integrated into the germ line and then passed onto the off ýÿ spring F3 where it was expressed specifically in the liver (Hew et al., 1995).

The introduction of AFPs to gold fish also increased their cold tolerance, to temperatures at which all the control fish died (12 h at 0o C; Wang et al., 1995). Similarly, injection or oral administration of AFP to juvenile milkfish or tilapia led to an increase in resistance to a 26 to 13o C. drop in temperature (Wu et al., 1998). The development of stocks harbouring this gene would be a major benefit in commercial aquaculture in counties where winter temperatures often border the physiological limits of these species.

The most promising tool for the future of transgenic fish production is undoubtedly in the development of the embryonic stem cell (ESC) technology. There cells are undifferentiated and remain totipotent so they can be manipulated in vitro and subsequently reintroduce into early embryos where they can contribute to the germ line of the host. This would facilitate the genes to be stably introduced or deleted (Melamed et al., 2002). Although significant progress has been made in several laboratories around the world, there are numerous problems to be resolved before the successful commercialization of the transgenic brood stock for aquaculture.

To realise the full potential of the transgenic fish technology in aquaculture, several important scientific break ýÿ through are required. There include (i) more efficient technologies for mass gene transfer (ii) targeted gene transfer technologies such as embryonic stem cell gene transfer (iii) suitable promoters to direct the expression of transgenes at optimal levels during the desired developmental stages (iv) identified genes of desireable traits for aquaculture and other applications (v) informations on the physiological, nutritional, immunological and environmental factors that maximize the performance of the transgenics of the transgenics and (vi) safety and environmental impacts of transgenic fish.

Chromosome Engineering

Chromosome sex manipulation techniques to induce polyploidy (triploidy and tetraploidy) and uniparental chromosome inheritance (gynogenesis and androgenesis) have been applied extensively in cultured fish species (Pandian and Koteeswaran, 1998; Lakra and Das, 1998). These techniques are important in the improvement of fish breeding as they provide a rapid approach for gonadal sterilization, sex control improvement of hybrid viability and clonation. Most vertebrates are diploid meaning that they possess two complete chromosome sets in their somatic cells. Polyploidy individuals possess on or more additional chromosome sets, bringing the total to three in triploids, four in tetraploids and so on.

Induced triploidy is widely accepted as the most effective method for producing sterile fish for aquaculture and fisheries management. The methods used to induce triploids and other types of chromosome set manipulations in fishes and the applications of these biotechnologies to aquaculture and fisheries management are well described (Purdom, 1983; Chourrout, 1987; Thorgaard, 1983; Pandian and Koteeswan, 1998). Tetraploid breeding lines are of potential benefit to aquaculture, by providing a convenient way to produce large numbers of sterile triploid fish through simple interploidy crosses between tetraploids and diploids (Chourrout et al., 1986; Guo et al., 1996). Although tetraploidy has been induced in many finfish species, the viability of tetraploids was low in most instances (Rothbard et al., 1997).

In teleosts, technique for inducing sterility include exogenous hormone treatment (Hunter and Donaldson, 1983) and triploidy induction (Thorgaard, 1983). The use of hormone treatements, however could be limited by governmental regulation and a lack of consumer acceptance of hormone treated fish products. Triploidy can be induced by exposing eggs to physical or chemical treatment shortly after fertilization to inhibit extrusion of the second polar body (For reviews see

Purdom, 1983; Thorgaard, 1983 and Ihssen et al., 1990) triploid fish are expected to be sterile because of the failure of homologous chromosomes to synapse correctly during the first meiotic division. Methods of triploidy induction in clued exposing fertilized eggs to temperature shock (hot or cold), hydrostatic pressure shock or chemicals such an colchicines, cytochalasin-B or nitrous oxide. Triploid can also be produced by crossing teraploids and diploids. Tetraploid induction involves fertilizing eggs with normal sperm and exposing the diploid sygote for physical or chemical treatment to suppress the first mitotic division. Gynogenesis is the process of animal development with exclusive maternal inheritance.

The production of gynogenetic individuals is of particular interest to fish breeders because a high level of inbreeding can be induced in single generation. Gynogenesis may also be used to produce all ýÿ female populations in species with female homogamety and to reveal the sex determination mechanisms in fish. It is convenient to use all female gynogenetic progenies (instead of normal bisexual progenies) for sex inversion experiments. Methodologies combining use of induced gynogenesis with hormonal sex inversion have been developed for several aquaculture species (Gomelsky et al., 2000). Androgenesis is the process by which would have commercial application in aquaculture. It can also be used in generating homozygous lines of fish and in the recovery of lost genotypes from the crypreserved sperms. Androgenetic individuals have been produced in a few species of cyprinids, cichlids and salmonids (Bongers et al., 1994).

Biotechnology and Fish Health Management

Disease problem area major constraint for development of aquaculture. Biotechnological tools such as molecular diagnostic methods, use of vaccines and immunostimulants are gaining popularity for improving the disease resistance in fish and shellish species world over for viral diseases, avoidance of the pathogen in very important. In this context there is a need to rapid method for detection of the pathogen. Biotechnological tools such as gene probes and polymerase chain reaction (PCR) are showing great potential in this area. Gene probes and PCR based diagnostic methods have developed for a number of pathogens affecting fish and shrimp (karunasagar ,1999). In case of finfish aquaculture, number of vaccine against bacteria and viruses have been developed. Some of these have been conventional vaccines consisting of killed microorgansism but new generation of vaccine consisting of protein subunit vaccine genetically engineered organism and DNA vaccine are currently under development.

In the vertebrate system, immunization against disease is a common strategy. However the immune system of shrimp is rather poorly developed, biotechnological tools are helpful for development of molecule, which can stimulate this immune system of shrimp. Recent studies have shown that the non specific defence system can be stimulated using, microbial product such as lipopolysacharides, peptidoglycans or glucans (itami et al 1998). Among the immunostimulants known to be effective in fish glucan and levamisole enhance phagocytic activities and specific antibody responses (Sakai, 1999).

Cryopreservation of Gametes or Gene Banking

Cryopreservation is a technique, which involve long-term preservation and storage of biological material at a very low temperature usually at -196 C ,the temperature of liquid nitrogen. It is based on the principle that very low temperature tranquilize or immobilize the physiological and biochemical activities of cell, thereby making it possible to keep them viable for very long period.

The technology of cryopreservation of fish spermatozoa (milt) has been adopted for animal husbandary . The first success in preserving fish sperm at low temperature was reported by Blaxter (1953) who fertilizes Herring (*Clupea herengus)* eggs with frozen thawed semen .The spermatozoa of almost all cultivable fish species has now been cryopreserved (Lakra 1993) . Cryopreservation overcomes problems of male maturing before female, allow selective breeding and stock improvement and enables the conservation (Harvey ,1996). One of the emerging requirement for that can be used by breeders for evolving new strains. Most of the plant varieties that has been produced are based on the gene bank collections.

Aquatic gene bank however suffers from the fact that at present it is possible to cryopreserve only the male gametes of finfishes and there in no viable technique for finfish eggs and embryos.

However, the recent report on the freezing of shrimp embryos. However, the recent report on the freezing of shrimps embryos by subramoniam and newton (1993) and Diwan and kandaswami (1997) look promising. Therefore, it is essential that gene banking of cultivated and cultivable aquatic species be undertaken expeditiously.

Aquaculture Biotechnology

Biotechnology is the fusion of biology and technology and refers to the application of biological techniques in product research and development. It involves the use of molecules (synthetic or physiological), biological materials (tissues, cells, genes) or processes derived from living organisms, to create new and practical applications for agriculture, medicine, industry and society.

What are its Applications for Reproductive Control in Fish Production?

The control of seasonal production and reproduction in farm animals have become major research goals. The applications of biotechnology to fish farming and ornamental fish production are numerous and valuable in both economic (food production, aquarium trade) and environmental terms (conservation of natural biodiversity for endangered species and protection of natural biodiversity from escapee domesticated strains). With the growing demand for fish products, biotechnology can help in the development of high quality, economical produce, thereby reducing pressure on natural populations.

In terms of fish reproduction for aquaculture, this means development of methods for the production of good quality gametes, independent of season, producing progeny with specific desired characteristics (good growth rate, flesh quality etc), meeting consumer demand. This technology should also contribute to the preservation and dissemination of genetic improvements.

Biotechnology may be used to:

- Control the sex of fish (e.g. produce single-sex populations)
- Delay the age of first sexual maturation (puberty)
- Control the reproductive cycle (in terms of gametogenesis and spawning)
- Support the genetic improvement of stocks, to produce strains more resistant to disease, displaying better growth rates (and food conversion ratios), more tolerant to stress and temperature changes, providing a product with an improved nutritional value, taste, texture and appearance.
- Improve management of gametes (short and long-term storage, transportation, development of media for in-vitro fertilisation)
- Conserve endangered or economically important strains via selective breeding programmes and management of captive broodstock (spermatogonial transplantation, cryopreservation of gametes)
- Produce sterile farmed animals to prevent the impact of escapees on the local biodiversity
- Assist in providing the aquarium trade with ornamental fish

Which Methods are Currently Practiced?

In-vitro fertilisation: This is the fertilisation of fish oocytes with manually/artificially collected sperm and ovules. Since the 1990's, research has permitted the development of several media (for dilution, to prevent spermatozoa activation, to preserve gamete fertilizing capability, for in-vitro gamete maturation...), thereby the success rates of the fertilisation process.

A dilution medium for washing oocytes, Ova-fish (IMV technologies), prior to in-vitro fertilisation, is currently available, which enables an in-vitro hatching rate success of 60-70% (for rainbow trout). Ovafish solution removes debris from the eggs, such as blood, urine and mucus, which may otherwise affect fertilisation success. Another medium, Actifish (IMV technologies), has been developed to improve the fertilising capability of fish sperm. Actifish is a fertilisation medium that can be used for fresh, conserved or frozen semence. Another medium, MaturFish (IMV technologies), has also been developed to promote sperm maturation. All these media have been specifically developed for the trout, but other examples exist for numerous species, including those of marine origin.

Long and Short Term Preservation of Gametes

Cryopreservation enables the long-term preservation of sperm of individuals of interest (disease resistant phenotypes etc.), for a virtually unlimited period of time.

This is useful for many reasons, for example: to develop breeding programmes independent of maturation period; to market standard quality sperm; to provide a year round supply of gametes; and to preserve a genetic stock of both commercially important strains as well as those species threatened by extinction in the wild. Research has permitted the development of several dilution mediums to protect sperm during freezing. However, freezing technology has very limited uses.

Oocytes cannot currently be frozen without losing their viability. However, research is in progress on the vernalisation of oocytes, this is very low temperature storage, without freezing, and could preserve the oocytes for several months.

Different media and temperatures have also been investigated for the short-term storage of both eggs and sperm. For example, studies have shown that trout eggs can be stored at 12°C for 3 days in coelomic fluid with no detrimental effect on egg quality.

Sex Control

The control of fish sex could be useful where one sex displays advantageous characteristics, such as larger adult size, production of high-value caviar (sturgeon), faster growth rate, or higher age at first sexual maturation (thereby avoiding degradation of flesh quality and certain bacterial infections). Sex control is currently only practiced commercially for salmonids, tilapia and hirame. Monosex populations of the most advantageous sex may be produced either by direct sex control via steroid treatment of the fish destined for consumption (masculinisation by administration of androgens; feminisation by administration of estrogens); or by genetic control and steroid treatment of broodstock (indirect hormonal treatment, gynogenesis, androgenesis, hybridisation); or by control of external factors (temperature, density etc.).

Genetic sex regulation may be achieved by:

- Crossing sex-reversed adult broodstock (administering androgens to produce "neo" males and estrogens to produce "neo" females; a "neo" fish is one which is genetically male or female, but phenotypically of the opposite sex) with normal males or females to produce single-sex progeny. Within Europe, only indirect androgen treatment of broodstock is permitted (i.e. production of all female progeny). This has obvious environmental advantages over the hormonal treatment of whole populations of pre-differentiated fish.
- Gynogenesis, or the fertilisation of oocytes with inactivated sperm (irradiated) with normal sperm, to eliminate the female chromosomes and produce all female progeny when the genotype is XX/XY. This method is more successful than androgenesis, but not widely used commercially.
- Androgenesis, or the fertilisation of inactivated oocytes (irradiated) with normal sperm, to eliminate the female chromosomes and produce all male progeny when the genotype is ZZ/ZW. In practice this method is not very successful and therefore not used commercially.
- Hybridisation of two different species from the same genera, producing single-sex progeny. However, the resulting offspring do not always display a satisfactory growth rate.

(Note: control of external factors, such as temperature for some gonochoristic fish - which do not change sex during their lifetime, such as tilapia or seabass; and social stimuli for hermaphrodites, such as seabream - which change sex during their lifetime, is a "gentle" method for controlling sex, but it is not yet commercially viable as it is not sufficiently reliable).

Control & Synchronisation of Spawning

The spawning season may be controlled by the manipulation of species-dependent environmental factors, such as photoperiod and temperature. These methods are routinely used to produce eggs in salmonid and some marine species such as seabass, turbot and seabream.

Spawning events may also be induced and spermiation enhanced by administration of gonadotropin releasing hormone analogs (GnRHa or LHRHa), hcG (human chorionic gonadotropin) or fish pituitary extracts. For example, within Europe, a synthetic GnRH analog, GonazonÒ (Intervet), is officially approved as an inducer of ovulation, showing good success rates in salmonids for both ovulation and subsequent fertilisation. In some other species, such as cyprinids, more potent inducers are used (hCG or Carp Pituitary Extract).

Other speciality pharmaceutical products such as OvaprimÒ have been developed, which combine a GnRH analog with domperidone, a potent inhibitor of the dopaminergic system. Fish farmers may also control production by speeding up or slowing down embryonic development by temperature and / or feeding practices.

Control of Puberty

Puberty may also be delayed by using different photoperiod regimes, for example in Atlantic salmon production. Puberty may also be completely eliminated, at least in females, by induction of sterility by triploidy. The application of antireceptor vaccinations, which have specific antagonistic effects of gametogenesis, is also in development.

Control of Fertility

The production of reproductively sterile fish (triploids) is very useful in responsible fish farm management and limits the genetic risk associated with the escape of domesticated fish into the wild. These farmed populations cannot interbreed with wild populations and therefore do not pose a threat to the natural biodiversity. In some species sterile fish reach a larger size without the complications of sexual maturity and the consequent reduction in flesh quality, which is desirable in sectors of the fish farming industry where large fish are most valuable. Generally, triploids grow more slowly than diploids. Triploidy is also not 100% successful (98% success rate), and as it is not feasible to test individual fish to check whether or not they are triploid, there will always be a certain number of individuals within the stock that are not triploid and are thus capable of reproducing. It should also be remembered that if the aim is to produce small sized fish (e.g. up to 200g), sexual maturity is not a problem as the fish will not attain maturity before they have reached their marketable size.

All female triploid offspring may be produced by the indirect use of externally administered hormones and by pressure / temperature treatment of eggs for a few minutes, post fertilisation. Within the European Union, it is forbidden to market hormone-treated fish for human consumption. Methyltestosterone may be used to sex-reverse broodstock females (XX) to neo-males (only sperm containing the X chromosome). These neo-males (XX) are then crossed with normal females (XX).

Induction of triploidy in the resulting zygotes will produce all-female triploid offspring (XXX). This is currently practiced on rainbow trout and perch. However, current research focuses on the development of alternative, more environmentally sensitive methods that are also species specific. Female triploids are always sterile, they do not undergo gonad development; however triploid males are only functionally sterile, they develop gonads and undergo puberty, but their sperm does not give rise to viable progeny.

Development of Gene Transfer Technologies for Fish Reproduction

Gene transfer technology has been applied to aquaculture species to obtain sterile fish by preventing the expression of GnRH. Further information can be obtained from the following papers:

- Wei Hu et al. 2007. Antisense for gonadotropin-releasing hormone reduces gonadotropin synthesis and gonadal development in transgenic common carp (Cyprinus carpio). Aquaculture 271: 498-506.

- Uzbekova, S. et al. 2000. Transgenic rainbow trout expressed sGnRH-antisense RNA under the control of sGnRH promoter of Atlantic salmon. J. Mol. Endocrinol. 25 (3): 337-350.

Principal Further Research Requirements

- Optimisation of species-specific hormonal treatments, in terms of the development of rapidly metabolised molecules that do not release biologically active catabolites into the environment (e.g. possible use of recombinant fish gonadotrophins for inducing gonadal development).
- Development of alternative methods to exogenous hormones based on genetic selection, environmental variable manipulation (e.g. photoperiod and temperature) and gene transfer technology, for fertility and sexual cycle control.
- Genetic basis of reproduction, for selection criteria / gene markers for broodstock companies
- Improved knowledge on important steps in fish reproductive cycles (sex control, puberty, gametogenesis and spawning)
- Improved knowledge on the determinants of gamete quality (e.g. surrogate broodstock technology)
- Research into reducing the impact of fish farms on the environment (in terms of the genetic integrity of the wild population, the discharge of hormones and antbiotics into the environment etc.)
- In terms of cryopreservation, a better knowledge of sperm membrane biology is needed. An improved understanding is required of the mechanisms regulating tolerance to freezing (for sperm and oocytes), as well as the formulation of improved protocols.
- Conservation of genetic resources and regeneration of populations of valuable animals or endangered species (e.g. spermatogonial transplantation).
- Development of generic methods for the domestication of new species impact of Science and Technology on Indian

Fisheries Sector

Fisheries sector, a sunrise sector in India, has recorded a faster growth than that of crop and livestock sectors. The sector contributes to the livelihood of a large section of economically-underprivileged population of the country.

It has been recognised as a powerful income and employment generator as it stimulates growth of a number of subsidiary industries and is a source of cheap and nutritious food besides being a foreign exchange earner.

With the changing consumption pattern, emerging market forces and technological developments, it has assumed added importance in India.

It is undergoing rapid transformation and the policy support, production strategies, public investment in infrastructure, and research and extension for fisheries have significantly contributed to the increased fish production.

Particularly, after the mid-1980s, the development of carp polyculture technology has completely transformed this traditional backyard activity into a booming commercial enterprise.

Fish Production in the World and in India

Approximately 50 million people worldwide depend on fishing for all or most of their family earnings, while another 150 million depend on fish processing and the fleet servicing industry. More than 10 million work on 2.5 million small-scale fishing vehicles and account for 50% of the world's catch (FAO 2001).

Fish production in the world rose from 23.50 million tonnes in 1950-51 to 140.48 million tonnes in 2003-04. Correspondingly, fish production in India has touched 6.40 million tonnes in 2003-04 from a mere 0.75 million tonnes in 1950-1951 (Table 1). The total fish production of our country stood at 6.87 million tonnes in 2006-07.

The share of India in global fish production has grown gradually from about 2.66% during the 1960s and 1970s to 4.56% in 2003-04.

Table: *Fish Production in World and India, 1950-51 to 2003-04 (Million tonnes)*

Year	World	% change	India	% change	India's share
1950-51	23.50	-	0.75	-	3.19
1960-61	43.60	85.53	1.16	54.67	2.66
1970-71	66.20	51.83	1.76	51.72	2.66
1980-81	72.30	9.21	2.44	38.64	3.37
1990-91	98.26	35.91	3.84	57.38	3.91
2000-01	129.00	32.35	5.66	47.40	4.39
2003-04	140.48	8.90	6.40	13.07	4.56
2006-07*	--	--	6.87	--	--

Source:

1. ICAR-ICLARM Project on 'Strategies and Options for Increasing and Sustaining Fisheries and Aquaculture Production to Benefit Poor Households in India, 2004 and Economic survey 2006-07, GOI
2. *Annual Report 2006-07 Provisional Estimates

Table1 shows that growth in fish production in India has been at a faster rate than in the world; mainly due to increasing contribution from inland fisheries. It can also be seen in Figures that in the pre-WTO period of 1990-91, the share of India's fisheries was 3.90, which rose to 4.56 in the post-WTO period of 2003-04. Overall, the share of developing world in the total world fish production increased from 43% in 1973 to about 73% in 1997, which has been mainly due to the increasing contribution from countries like China and India (Delgado *et al*, 2003).

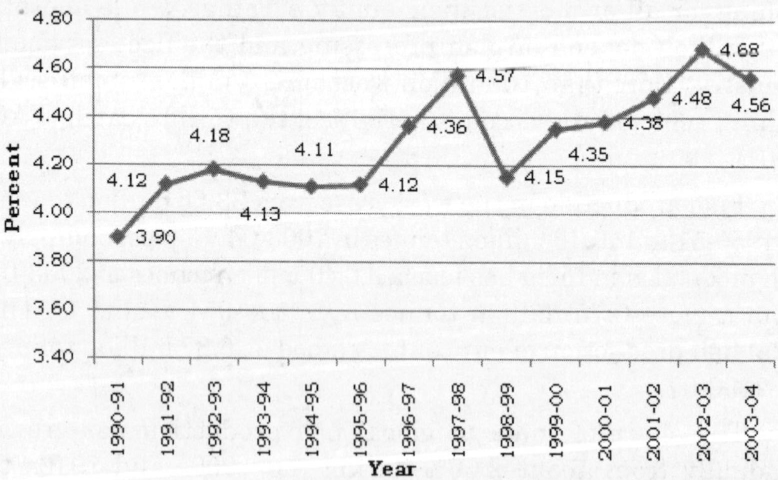

Figure: *India's Share in World Fish Production*

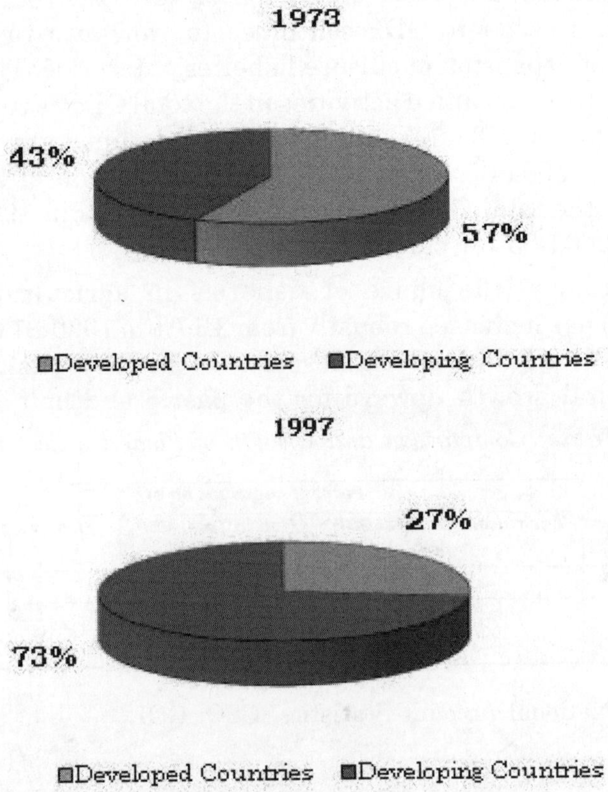

***Figure:** Changing Share of Developing Countries in the Production of Fish for Food, 1973 and 1997*

Contribution to Indian Economy and Prospects of Fisheries Growth

With fisheries sector comprising marine fisheries, freshwater and brackish water aquaculture and inland fisheries consisting of tanks and reservoir, the potentiality of this sector as a whole remains to be fully tapped and it remains a sector of much promise. The fisheries sector in particular is more complex enterprise that functions under integrated network of natural resources, other enterprises with forward and backward linkages with fisheries and other socio-political variables. The major functions of fisheries enterprises, viz. production, transportation, storage and processing involve value addition from labour, capital and management, which significantly influence the rapid economic development of the country.

In the last 25 years, unlike agriculture, the contribution of fisheries sector to GDP continued to grow at a rapid pace because of expansion of culture fisheries enterprise. The share of agriculture and allied activities in the total GDP is constantly declining. It was 34.69% in 1980-81 and declined gradually to 17.62% in 2004-05. In contrast, the contribution of fisheries sector to the total GDP has gone up from 0.75% in 1980-81 to 1.04 in 2004-05 (at current prices).

Similarly, the share of fisheries in agriculture GDP (AgGDP) has increased robustly from 2.17% in 1980-81 to 5.93% in 2004-05 (Table 2). This sector is in fact pushing the agricultural growth upward for the past 5 and half decades.

Table: *Comparison and Growth of Fisheries Sector*

Period	***Percent contribution of***		
	Agriculture to Total GDP	***Fish to Total GDP***	***Fish to Ag GDP***
1980-81	34.69	0.75	2.17
1990-91	28.42	0.96	3.37
2000-01	22.26	1.18	5.32
2004-05	17.62	1.04	5.93

Source: National Account Statistics, CSO, GOI.

Chapter 3

Fish Production: Structure and Composition

The fish production in India witnessed a spectacular growth since independence. It rose from a mere 0.75 million tonnes in 1950-51 to over 6.57 million tonnes in 2005-06 (Table 3). In the initial years, marine sector used to contribute more to total fish production then inland sector. In 1950-51, marine production contributed about 71.01%, which fell gradually to 42.77% in 2005-06, while inland sector started contributing from 28.99 in 1950-51 to about 57.23% in 2005-06. In fact, by the year 2000, its share crossed 50% and continues to improve its share further in the coming years.

Table: Fish Production by Source in India

Year	Fish production (Mt)		
	Marine	Inland	Total
1950-51	0.534	0.218	0.752
1960-61	0.880	0.280	1.160
1970-71	1.086	0.670	1.756
1980-81	1.555	0.887	2.442
1990-91	2.300	1.536	3.836
2000-01	2.811	2.845	5.656
2003-04	2.941	3.458	6.399
2004-05	2.779	3.526	6.305
2005-06	2.816	3.756	6.572

Expansion of fleet capacity, technological innovation, and increases in investment all led to explosive growth in the exploitation of marine fisheries through the 1960s, 1970s and

1980s. But from the late 90s onwards, the marine fisheries production reached a plateau and it seems that it can register only marginal increase in the near future. With most wild fisheries near maximum sustainable exploitation levels, capture fisheries will most likely grow slowly.

On the other hand, inland fish production was on constant rise. The inland fisheries include both capture and culture fisheries. The capture fisheries have been the major sources of inland fish production till mid-1980s.

But the fish production from natural waters like rivers, lakes, etc. followed a declining trend, primarily, due to proliferation of water control structures such as dams and habitat degradation due to human interference and settlement near and surrounding water bodies (Katiha and Bhatta, 2002). The depleting resources, energy crisis and resultant high cost of fishing etc. have led to increased realisation of the potential and versatility of aquaculture as a sustainable and cost-effective alternative to capture fisheries.

Trends and Current Patterns of Fish Production

To assess the trend of fish production in the country, the last 25 years data (from 1980-81 to 2005-06) was used. Both linear and non-linear trend analysis was tried and the best equation fitted. The total fish production in India has been following a linear trend and it is likely proceed further in the same direction.

The trend pattern of marine and inland fish production revealed that while the marine sector's production was increasing at a decreasing rate, the inland sector's production was increasing at an increasing rate, the possible reasons for which are that inland aquaculture activities are gaining much importance in some of the states like Uttar Pradesh, Andhra Pradesh, Punjab etc. in recent years. The trend equations for both marine and inland fish production, as well as for total fish production are given below:

Marine: $Y = -2.7588X + 143.4X + 1061.3$

Inland: $Y = 2.9078X + 38.083X + 824.85$

Total: $Y = 185.5X + 1867.4$

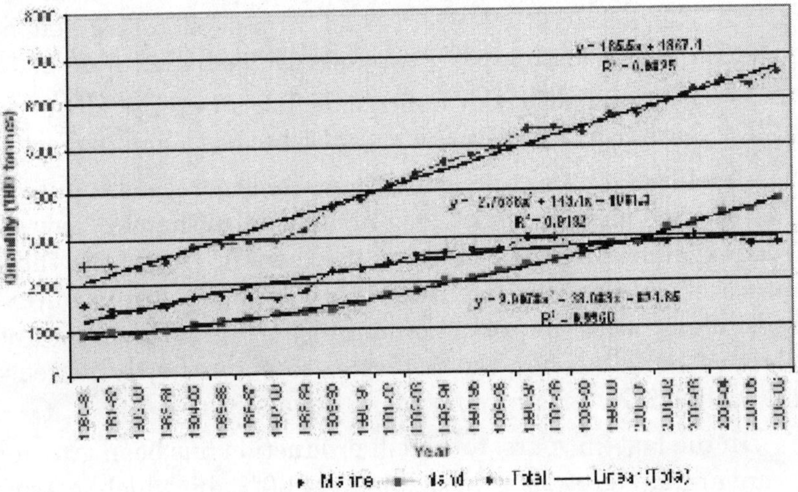

Figure: *Trend in Fish Production, 1980-81 to 2005-06*

Based on the trend observed in the last 25 years, assuming that the same scenario would prevail in both public and private investment pattern in the sector, the fish production of the country would reach 7.46, 8.42 and 9.38 million tonnes in 2010, 2015 and 2020 respectively (Table 4).

Table: *Fish Production in 2020*

Year	***(in million tones)***		
	Marine	***Inland***	***Total***
2010	2.88 (38.61)	4.58 (61.37)	7.46
2015	2.70 (32.07)	5.72 (67.33)	8.42
2020	2.38 (25.37)	7.00 (74.63)	9.38

Note: Parentheses indicates percentage to total

In the coming decades, aquaculture will likely be the greatest source of increased fish production, as fish farmers expand the water surface area under cultivation and increase yields per unit of area cultivated. But the sector must overcome several major challenges, if it has to sustain the rapid growth of the past 25 years:

- Aquaculture has to face competition from others users of land and water, as these resources would become more scarce in future (Rosegrant *et al*, 2002)
- Marine sector would face energy crisis, as it would need more fossil fuels for further expansion

- The aquaculture production would be restrained by diseases as the sector expands (Subasinghe, *et al*, 2001)
- The availability of fish meal, fish oil and wild caught trash fish as feed inputs may also become a limiting factor.

Yield increases can come either from increased inputs or greater efficiency of inputs. It is likely that in the next several decades, aquaculture production will benefit from both these sources of yield growth. Greater use of compounded aquaculture feeds along with improvements in rearing technology and selective breeding has the potential to significantly increase the productivity of many forms of aquaculture.

In the last 25 years, total fish production has been growing at an annual growth rate of about 4.60%, in which marine sector was growing at a rate of 3.24 and inland sector was growing at a rate of 6.20%.

***Table:** Compound Growth Rate in Fish Production, 1980-81 to 2005-06*

Year	Marine	Inland	Total
1980-81 to 1989-90	3.80	5.28	4.39
1990-91 to 1999-00	2.33	6.55	4.13
2000-01 to 2005-06	-0.21	5.37	2.75
1980-81 to 2005-06	3.24	6.20	4.60
1980-81 to 1980-91 (Pre-WTO)	4.35	5.43	4.78
1991-92 to 2005-06 (Post-WTO)	0.84	5.71	3.18

Overall, inland sector fared better in all the periods, viz. 1980-81 to 1989-90, 1990-91 to 1999-00 and 2000-01 to 2005-06. There seems to be a slower pace in growth of this sector in the recent times. In contrast, marine sector witnessed a negative growth rate in the period 2000-01 and 2005-06, which indicates the exhaustion of marine resources especially in the in-shore and near-shore waters, where maximum harvesting has been carried out. About 90% of the present production from the marine sector is within a depth range of up to 50-70 m and the remaining 10% from depths extending upto 200m. While 93% of the production is contributed by artisanals and motorized sectors, the remaining 7% is contributed by deep sea fishing fleets confining their operation mainly to the shrimp grounds in the upper east coast. Hence, in order to enhance and sustain the contribution from this sector, we need to go into the deep sea for targeting its untapped potential. This requires enhanced investment in mechanized vessels, capacity strengthening of artisanals and probably a proper institutional structure to share the benefits.

The growth rates in pre and post-WTO periods were also estimated. It is noticed that the pre-WTO period witnessed an impressive growth rate of about 4.78 as compared to post WTO period (3.18). This trend was mainly due to the marine sector, which is understandable by the fact that the country's fish export basket was dominated by marine species and buoyance of marine export might have propelled the growth of marine catch, and vice versa. The post-WTO period imposed many quality regulations in terms of SPS measures on developing countries like India, which couldn't create huge investment in the infrastructures required to produce export-quality marine fisheries products that are acceptable to our trading partners, especially EU, USA and Japan. In contrast to the marine sector, inland sector continued to grow better in the post-WTO period also, which is possibly because of enhanced public and private investment in inland fisheries sector especially though different development programmes and research by the Government of India since IV plan, that started delivering results continuously.

Figure clearly depicts the dominance of inland fisheries sector in the contribution to total fish of the country, particularly from 1990s.

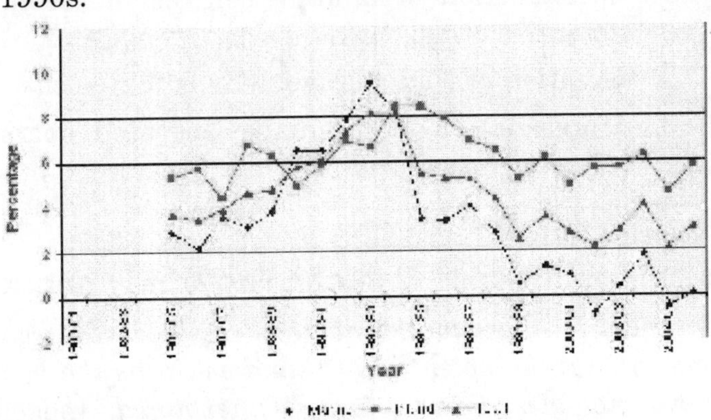

Figure: *Annual Growth Rate (%) in Fish Production, 1980-81 to 2005-06*

Drivers for Future Growth

To sustain this growth of the sector in general; technology, infrastructure and market would play a major role apart from enhanced investment in research and development. Technology had been the main factor responsible for the phenomenal growth of aquaculture, particularly after the advent of carp polyculture and composite fish culture in the late 70s.

Similarly, major investments on infrastructure such as construction of mini harbors, jetties, landing centres, introduction of trawlers and mechanized vessels, supply of nets, etc. led to increased catch and contribution from capture fisheries sector. However, market has not been able to play the role of a major driver for the growth of the sector so far. To un-tap the potential of the sector, market should take the lead in furthering the growth, especially in the emergence of aquaculture sector. Some of the illustrative aspects under each major driving force are given below for future attention:

1. Technology
    1. Quality seed production
    2. Selective breeding of carps
    3. Formulation of low cost feed materials using locally available ingredients
    4. Fabrication of nets for targeted fishing
2. Infrastructure
    1. Construction of landing centres in second tier potential costal towns
    2. Creation of cold storage near landing centres
    3. Up-gradation of manually operated boats into outboard motorized ones
    4. Supply of ice box

Opportunities in Marine Biotechnology and Aquaculture

The oceans offer abundant resources for research and development, yet the potential of this domain as the basis for new biotechnologies remains largely unexplored. Indeed, the vast majority of marine organisms (primarily microorganisms) have yet to be identified. Even for known organisms, there is insufficient knowledge to permit their intelligent management and application. Oceanic organisms are of enormous scientific interest, for two major reasons. First, they constitute a major share of the Earth's biological resources. Second, marine organisms often possess unique structures, metabolic pathways, reproductive systems, and sensory and defence mechanisms because they have adapted to extreme environments ranging from the cold polar seas at -2° C to the great pressures of the ocean floor, where hydrothermal fluids spew forth. Most major classes of the Earth's organisms are primarily or exclusively marine, so the oceans represent a source of unique genetic information.

The oceans therefore offer abundant resources for research and development. Yet the potential of this domain as the basis for new biotechnologies remains largely unexplored. Indeed, the vast majority of marine organisms (primarily microorganisms) have yet to be identified. Even for known organisms, there is insufficient knowledge to permit their intelligent management and application.

Interest in marine biotechnology has been growing in recent years, (2) By contrast, the U.S. Government invested less than $44 million on marine biotechnology R&D in fiscal year 1992, (4). Only about 1 or 2 percent of the total Federal investment in biotechnology has been devoted to marine biotechnology and aquaculture. Even so, promising results are emerging from these activities. Additional Federal support of research in key areas of marine biotechnology and aquaculture will generate both new fundamental knowledge and advanced technologies for producing new pharmaceuticals, biomaterials, and other products; developing and improving bioremediation and bioprocessing; enhancing cultivation of aquatic species; and expanding understanding of biological processes in the oceans and their role in global change.

This chapter provides an overview of recent biotechnology applications in marine science and aquaculture and identifies research priorities and specific opportunities that offer potential for significant advances. The chapter emphasizes development of salt-water resources but includes biotechnology applications involving fresh water, including thermal springs.

The five research priorities are:

- Develop a fundamental understanding of the genetic, nutritional, and environmental factors that control the production of primary and secondary metabolites in marine organisms, as a basis for developing new and improved products.
- Identify bioactive compounds and determine their mechanisms of action and natural function, to provide models for new lines of selectively active materials for application in medicine and the chemical industry.
- Develop bioremediation strategies for application in the world's coastal oceans, where multiple uses — including wastewater disposal, recreation, fishing, and aquaculture — demand prevention and remediation of pollution; and develop bioprocessing strategies for improving sustainable industrial processes.
- Use the tools of modern biotechnology to improve the health, reproduction, development, growth, and overall well- being of cultivated aquatic organisms; and promote the interdisciplinary development of environmentally sensitive, sustainable systems that will enable significant commercialization of aquaculture.
- Improve understanding of microbial physiology, genetics, biochemistry, and ecology in order to provide model systems for research and production systems for commerce, and to contribute to understanding and conservation of the seas.

New and Improved Products from The Seas

PRIORITY: Develop a fundamental understanding of the genetic, nutritional, and environmental factors that control the production of primary and secondary metabolites in marine organisms, as a basis for developing new and improved products.

PRIORITY: Identify bioactive compounds and determine their mechanisms of action and natural function, to provide models for new lines of selectively active materials for application in medicine and the chemical industry. Most major groups of living organisms are primarily or exclusively marine. Tropical marine environments harbor an especially wide diversity of animals and plants. Many marine organisms are sessile and must employ sophisticated methods to compete for a place to anchor. This characteristic is reflected in part by a metabolism that produces enormously diverse bioactive products — many with no terrestrial counterparts.

Recent research has uncovered unicellular and multicellular microorganisms that are unique to the marine world. Indeed, "... it seems clear that marine bacteria are emerging as a significant chemical resource".(6) and the amounts and characteristics of chemicals they produce.

Federal support for research in this area is essential, both to answer the fundamental questions and to assure that the resulting knowledge is translated into sustainable technologies. While it will be vital to cultivate marine microorganisms that produce novel products, the alternative approach of transferring genes of interest into non-marine microorganisms also should be investigated. For example, the capability to produce a marine polysaccharide — a complex molecule that could be useful as a food additive or a water-resistant adhesive — could be transferred to an easily grown bacterium (e.g., *E. coli* or *Bacillus subtilis*). This approach might be more effective in some cases than would cultivating the marine organism or recreating artificially the long and complicated production pathway for the polysaccharide.

Pharmaceuticals

Many bioactive substances from the marine environment already have been isolated and characterized, several with great promise for the treatment of human diseases. The compound manoalide from a Pacific sponge, for example, has spawned more than 300 chemical analogs, with a significant number of these going on to clinical trials as anti-inflammatory agents.

To promote sustainable technology development, the Federal Government should support research leading to new methods for discovery. One approach would be to focus on learning about natural functions, regulation, and production of substances generated by marine organisms, in order to identify potentially important agents. Refined test systems should be developed in order to identify selective agents produced by marine organisms.

Rapidly developing assay technology can facilitate exploration of the bioactivity of newly discovered compounds. These assay methods, which employ specific receptors for known physiological agents, require only minute amounts of a test substance and can be automated. (Traditional chemical tests require considerable amounts of test material and labourious measurement processes.) The new tools will make it feasible to test newly discovered compounds rapidly by the hundreds, for a wide spectrum of biological activities.

To date, exploitation of natural agents from the sea has been hindered by problems with limited or sporadic distribution and production. Much more research must be conducted to determine what seasonal factors and life cycle or reproduction states are linked with natural production of an agent. Factors influencing production may include diet, physical and chemical conditions, distribution by phylogenetic affiliation, geographic location, water depth, or associations with symbiotic microorganisms. Knowledge of these factors will be important in developing methods for producing selected metabolites, either from whole organisms or *in vitro* from cell or tissue cultures of plants and animals. Many of these compounds are very large and complex molecules, requiring very elaborate biochemical processes; as a result, it can be difficult to synthesize them or clone all the genes for standard production through fermentation.

Enzymes

Enzymes produced by marine bacteria are important in biotechnology due to their range of unusual properties. Some are salt-resistant, a characteristic that is often advantageous in industrial processes. The extracellular proteases are of particular importance and can be used in detergents and industrial cleaning applications, such as in cleaning reverse-osmosis membranes. *Vibrio* species have been found to produce a variety of extracellular proteases. *Vibrio alginolyticus* produces six proteases, including an unusual detergent-resistant, alkaline serine exoprotease. This marine bacterium also produces collagenase, an enzyme with a variety of industrial and commercial applications, including the dispersion of cells in tissue culture studies.

Other research has demonstrated the presence in algae of unique haloperoxidases (enzymes catalyzing the incorporation of halogen into metabolites). These enzymes could become valuable products, because halogenation is an important process in the chemical industry. Japanese researchers have developed methods to induce a marine alga to produce large amounts of the enzyme superoxide dismutase, which is used in enormous quantities for a range of medical, cosmetic, and food applications.

An unusual group of marine microorganisms from which enzymes have been isolated are the hyperthermophilic archaea (previously called archaebacteria), (9) which can grow at temperatures over 100° C and therefore require enzyme systems that are stable at high temperatures. Archaea typically are found in extreme environments, such as hot springs, animal guts, hydrothermal vents, sewage sludge digesters, and hypersaline habitats, including the Great Salt Lake.

Thermostable enzymes offer distinct advantages, many still to be discovered, in research and industrial processes. Thermostable DNA-modifying enzymes, such as polymerases, ligases, and restriction endonucleases, already have important research and industrial applications. Hot springs in Yellowstone National Park provided the first archaeon (*Thermus aquaticus*) from which thermostable DNA polymerases were isolated. These novel enzymes (the *Taq*® polymerases) became the basis for the polymerase chain reaction (PCR), a useful technique for studying genetic material. In 1989, thermostable DNA polymerase was designated Molecule of the Year by Science magazine. Comparable enzymes continue to be discovered.

Most enzymes involved in the primary metabolic pathways of thermophilic bacteria and archaea are dramatically more thermostable than are their counterparts living at moderate temperatures. Expanded study of enzymes from thermophilic marine microorganisms will contribute to the understanding of mechanisms of enzyme thermostability and should enable the identification of enzymes suitable for industrial applications as well as modification of enzymes to enhance thermostability.

Biomolecular Materials

Recent research has demonstrated that marine biochemical processes can be exploited to produce new biomaterials. For example, a corporation in Chicago is commercializing a new class of biodegradable polymers modelled on natural substances that form the organic matrices of mollusk shells. Equally exciting are the mechanisms used by marine diatoms, coccolithophorids, mollusks, and other marine invertebrates to generate elaborate mineralized structures on a nanometre scale (less than a billionth of a metre in size).

Nanometre-scale structures can have unusual and useful properties. Research that will enhance understanding and allow engineering of the processes for creating these bioceramics promises to revolutionize the manufacturing of medical implants, automotive parts, electronic devices, protective coatings, and other novel products.

Biomaterials also hold promise for counteracting biofouling, which long has been recognised as an extensive and costly problem. Bacterial biofilms form slime layers that increase drag on moving ships, interfere with transfer on heat exchangers, block pipelines, and contribute to corrosion on metal surfaces. Bacterial and microalgal colonization of surfaces is accompanied by settlement of invertebrate larvae and algal spores, eventually leading to "hard fouling" and the need for costly cleaning.

The most effective anti-fouling coatings have utilised toxic chemicals, such as copper and organotins. There is an urgent need for non-toxic biofouling control strategies, due to heightened recognition of the impact that toxic coatings can have on the environment. Research is needed on the attachment mechanisms of marine organisms and the natural products they employ to prevent fouling of their own surfaces.

Molecular approaches to characterizing biofilm structure and development offer considerable potential for finding novel biofouling prevention strategies. It is now possible to determine the genes and pathways involved in regulation and synthesis of bacterial adhesive polymers. Considerable progress has been made in understanding the nature and expression of surface polymers produced by microorganisms such as the nitrogen-fixing *Rhizobium* species and the opportunistic pathogen *Pseudomonas aeruginosa*. Similar approaches can been applied to marine biofilm bacteria, to find the genetic determinants of adhesive production and the environmental factors that regulate synthesis.

Molecular biology techniques also can be applied to determine the basis of natural antifouling mechanisms. Many marine plants and animals remain free of attached bacteria, either because they produce repelling compounds or because their surface structure neutralizes bacterial adhesives. A product generated by the seagrass *Zostra marina* (eelgrass), for example, is an effective agent for preventing fouling by bacteria, algal spores, and a variety of hard-fouling barnacles and tube worms.(11) Molecular characterization of natural fouling resistance could provide new strategies for fouling control. Potential applications include prevention of fouling in industrial pipelines or heat exchangers, improved design of trickling filters or aquaculture circulation systems, and control of biofilm infections of medical implants and prosthetic devices.

Biomonitors

Marine organisms can provide the basis for development of biosensors, bio-indicators, and diagnostic devices for medicine, aquaculture, and environmental monitoring. One type of biosensor employs the enzymes responsible for bioluminescence. The *lux* genes, which encode these enzymes, have been cloned from marine bacteria such as *Vibrio fischeri* and transferred successfully to a variety of plants and other bacteria.

The *lux* genes typically are inserted into a gene sequence, or operon, that is functional only when stimulated by a defined environmental feature. The enzymes responsible for toluene degradation, for example, are synthesized only in the presence of toluene. When *lux* genes are inserted into a toluene operon, the engineered bacterium glows yellow-green in the presence of toluene. This genetically engineered system "reports" that biodegradation of a specific chemical, in this case toluene, is proceeding. Another type of biomonitor that holds great promise is the gene probe, which can be used to identify organisms that pose health hazards or may be useful in research. Specific gene probes can be employed, for example, to detect human pathogens in seafood and recreational waters; fish pathogens in aquaculture systems; microorganisms capable of mediating desired chemical transformations (e.g., toxic chemical degradation, CO2 assimilation, metal reduction); and specific fish stocks in fish migration and recruitment studies.

Biopesticides

Natural marine products have the potential to replace chemical pesticides and other agents used to maximize crop yields and growth. Continued Federal support for R&D in this area is likely to result in useful natural pesticides that would provide greater specificity and fewer harmful side effects than do conventional synthetic agents. Current U.S. expenditures for all pesticides amount to $47 billion annually; by the year 2000, biopesticides from marine and other sources are expected to capture an estimated 10 percent of this market.(12)

An example of a marine biopesticide in use today is Padan*TM*, which was developed from a bait worm's toxin known to ancient Japanese fishermen. This natural pesticide has demonstrated activity against larvae of the rice stem borer, the rice plant skipper, and the citrus leaf miner, among other pests. More recently, scientists in Montana discovered novel compounds in marine algae and marine sponges containing symbiotic microorganisms. These compounds promoted growth and stimulated germination and increased root and coleoptile lengths in test plants.

Several sponge and nudibranch species produce terpenes, a broad class of aromatic compounds used in solvents and perfumes and known to deter feeding by fish. Extracts derived from these same sponge and nudibranch species also demonstrated powerful insecticidal activity against two species, grasshoppers and the tobacco hornworm.(13)

Biomass for Energy Production

Approximately 40 percent of all primary energy production, or photosynthesis, occurs in the seas. In this process, oceanic plants (phytoplankton, seaweeds, seagrasses) take up carbon dioxide (CO_2) and, with light energy from the sun, convert it into organic carbon (primarily sugars) and oxygen. The oceans contain 50 times as much carbon dioxide as does the atmosphere, and it is estimated that primary production incorporates 35 gigatons (1 gigaton = 1 x 10*15* grams) of carbon into marine biomass annually. This abundant source of fuel for energy production has not been tapped commercially because it is not competitive with soybean meal and other easily harvested, traditional sources of biomass, and also because, regardless of the source, biomass is not competitive with other types of fuels.

The Federal Government should continue to support research on the use of biotechnology to enhance biomass production and utility. At least three general approaches are being explored.(14)

First, the enzyme that captures CO_2 for photosynthesis — ribulose bisphosphate carboxylase\oxygenase or "RUBISCO" — is relatively inefficient, so supercomputers are being used to verify structural information, and the enzyme is being redesigned to optimize its function. Second, the chemical composition of biomass can be altered to make it more suitable for particular applications. For example, marine microalgae are being genetically engineered to boost their lipid content, with the aim of providing a source of alternative fuels that is more economical than are conventional sources. Third, biotechnology is being used to convert biomass to ethanol and other alternative forms of energy and chemical feedstocks.

Aquaculture

PRIORITY: Use the tools of modern biotechnology to improve the health, reproduction, development, growth, and overall well-being of cultivated aquatic organisms; and promote the interdisciplinary development of environmentally sensitive, sustainable systems that will enable significant commercialization of aquaculture.

Aquaculture, which long has been practiced in Asia and is increasingly popular in the United States, Europe, and South America, will benefit tremendously from the use of new molecular tools and processes. With worldwide seafood demand projected to increase 70 percent in the next 35 years, and harvests from capture fisheries stable or declining, aquaculture will have to produce seven times as much seafood as it generates now to supply global demand by 2025.(17) The use of modern biotechnology to intervene in the rearing process and enhance production of aquatic species holds great potential not only to meet this demand, but also to improve U.S. competitiveness in aquaculture.

The U.S. aquaculture industry has grown rapidly in recent years. Farm gate receipts exceeded $800 million in 1992 — a fourfold increase since the early 1980s.(19)

The growth and international competitiveness of the U.S. aquaculture industry will be determined by the size of the resource investment in research and technology development. This investment should be made through a partnership of Federal and state agencies and the private sector. The Federal role is to provide leadership in supporting research to advance knowledge in important research areas and to facilitate the transfer of promising results and technologies to the private sector.

The major research issues in aquaculture are similar to those for other agricultural sectors, but the knowledge base for aquaculture is comparatively meager. Development of this knowledge is a particular challenge due to the diversity of cultured aquatic species and the systems for their production. Federal support for biotechnology research in this area will expand the knowledge base and yield significant dividends.

The application of biotechnology promises significant benefits to both producers and consumers of aquacultural products. The use of genetically enhanced organisms may improve production efficiency through improvements in growth rates, food conversion, disease resistance, and product quality and composition. The application of biotechnology to aquaculture also may help conserve wild species and genetic resources and provide unique models for biomedical research.

Enhancing Reproduction and Early Development

Biotechnology can be applied to enhance reproduction and early development of cultivated aquatic organisms. The resulting benefits could include year-round production of gametes and fry of economically valuable species and creation of new markets for specialized, genetically improved broodstock. Similarly, biotechnology may provide techniques for improving the reproductive success and survival of endangered species, thereby helping to preserve the diversity of life on Earth.

As a first step, research should be directed toward improving basic understanding of environmental, hormonal, biochemical, and genetic control of reproduction. More specifically, scientists must identify and understand the mechanisms of expression of genes involved in reproduction and development, improve technologies for cryo-preserving gametes and embryos, improve delivery systems for administration of natural and synthetic hormones, and enhance understanding of the pharmacokinetics of uptake and release of administered hormones.

Improving Health and Well-Being

Biotechnology offers substantial opportunities to improve the health and well-being of cultivated aquatic organisms. More than 50 diseases affect fish and shellfish cultured in the United States, causing losses of tens of millions of dollars annually. Biotechnology not only can improve the survival, growth, vigor, and well-being of cultivated stocks, but also can reduce disease transfer between cultivated and wild stocks. New products and market opportunities can be developed related to aquatic animal health and well-being.

The tools of molecular biology can provide a basic understanding of host immunity, resistance, and susceptibility to diseases and associated pathogens by furnishing information about life cycles and mechanisms of pathogenesis, antibiotic resistance, and disease transmission. Improved technologies must be developed for detecting and diagnosing pathogens and diseases and for enhancing the genetic basis of disease resistance, thereby reducing the need for antibiotics and other drugs. Potential products resulting from this research include gene therapy techniques; broodstock free of pathogens; safe, effective prophylactic agents, including immune modulators, antigens, and vaccines; safe, effective therapeutic agents; and improved systems for administering prophylactic and therapeutic agents.

Improving Quality and Value

The Federal Government has a responsibility to help ensure the safety and quality of food supplies, and biotechnology can and should be an invaluable tool in carrying out this mandate. Biotechnology can be employed to assess and improve the safety, freshness, colour, flavour, texture, taste, nutritional characteristics, and shelf life of aquacultural food products. In addition, practical technologies can be developed to detect and assay toxins, contaminants, and residues in seafood, and to reduce or eliminate contaminants. There are also opportunities to apply biotechnology in improving seafood processing. Research should be conducted to develop and improve technologies for all these applications.

Conserving Genetic Resources

The preservation and enhancement of biodiversity in natural systems is an important Federal priority. Therefore, the Federal Government should encourage and support programs to maintain and enhance biodiversity in aquatic systems through cultivation and stocking of aquatic species. Biotechnology can be employed in two ways to conserve genetic resources of aquatic species.

First, the tools of biotechnology can be used to identify and characterize important aquatic germplasm, including endangered species. Genomes of aquatic species can be analysed and characterized, and quantitative trait loci identified. Second, biotechnology can be applied to improve understanding of the molecular basis of gene regulation and expression as well as sex determination and thereby improve methods for defining species, stocks, and populations. Approaches include developing marker- assisted selection technologies, improving precision and efficiency of transgenic techniques, and improving technologies for the cryo-preservation of gametes and embryos. Ultimately, stocking certain areas with selected, cultivated species and strains could help maintain biodiversity in natural aquatic ecosystems.

Enhancing Biomedical Models

Aquaculture has important purposes other than food production. Because they often adapt to extreme environments, marine organisms can provide unique models for research on biological and physiological processes. Studies of the developmental, cellular, and molecular aspects of marine organisms as model systems will provide insights into the basis of disease mechanisms and pathogenesis in humans. By contrast, use of mammalian organisms as a basis for the development of some types of human disease models may be neither feasible nor cost effective.

Progress in this arena will require that sophisticated molecular biology technologies be adapted to marine organisms, in order to enhance understanding of their biological processes. For example, approaches for gene transfer into eggs have been developed for many terrestrial organisms, but not for most marine species. This technology is needed for analyses of gene regulatory systems and gene expression. In addition, methods need to be developed for culturing tissues from marine organisms. Cultured cell lines will provide opportunities for gene transfer and gene expression studies and enhance the usefulness of marine species as biomedical research models. This is an important research area deserving of Federal support.

Understanding and Conserving the Seas

PRIORITY: Improve understanding of microbial physiology, genetics, biochemistry, and ecology in order to provide model systems for research and production systems for commerce, and to contribute to understanding and conservation of the seas.

Scientists have a powerful new array of sampling devices and measuring instruments that will accelerate greatly the acquisition of knowledge about ocean resources and foster their wise use. These technologies include manned deep-sea submersibles, remotely operated vehicles, geosynchronous satellites, sophisticated acoustic measuring devices, pressure- retaining deep-sea samplers, geographic information systems, real-time flow cytometry, PCR and biomonitoring techniques, computerized databases, and other forms of information exchange and analysis. These tools should be exploited to accelerate the discovery of unknown marine microorganisms and to expand understanding of known varieties. Federal support for this research is essential, because only then will sufficient information be acquired to assure that practical applications will result. As new life forms and processes become known, and as understanding of them grows, marine biotechnology will make significant contributions to the nation's social and economic well-being.

Identifying Organisms and Their Niches

Biodiversity

Nowhere in the biosphere is biological diversity greater than in the seas, and the extent of this diversity becomes increasingly evident as scientists investigate new environments. In the 1980s, for example, giant tube worms (*Riftia*) were recovered from areas adjacent to deep-ocean thermal vents, and novel mussels (*Bathymodiolus*) that farm methanotrophic bacteria on their gill tissue were discovered around methane seeps in the Gulf of Mexico. Most newly described species have been and will continue to be microorganisms, although it is clear that new marine plants and animals also await discovery.

Less than 1 percent of the extant bacteria — marine and terrestrial — have been isolated and described.(20) The marine environment represents a particularly fertile source for new bacteria, as evidenced by the recent discovery of unusual "cold water" archaea 100-500 metres deep in the oceans.(21) These archaea comprise a high percentage of the total bacterial ribosomal RNA present in seawater samples, yet they have not been isolated in pure culture and described. Pursuit of this research may provide new understanding of oceanic processes and, once these archaea are cultivated, perhaps a novel source of products. Countless other marine bacteria also have yet to be cultured; advances in microbial culture technology will enable scientists to isolate, describe, and possibly exploit these organisms.

Viruses are another newly appreciated element of marine biological diversity. Electron microscopy and PCR technology have revealed abundant numbers of viral particles in seawater samples. Specific viruses that infect species of marine phytoplankton have been cultured from coastal as well as open ocean sites.(22) This discovery is exciting for at least two reasons.

Viral infection of higher forms of marine life almost certainly affects global ocean processes, such as photosynthesis. Marine viruses also will provide new materials for development of genetic and biotechnological tools that can be used to study and manipulate marine organisms. For example, marine viruses could be used to genetically engineer higher forms of marine life.

Marine Ecology

As a basis for developing new applications for marine products and processes, analysing global climate change, and improving fisheries management, scientists must build a fundamental understanding of marine organisms and their specific adaptations to and interactions with their environment. Recent developments in molecular techniques are rapidly expanding capabilities for research on marine ecology. Three research areas seem to be especially fertile and deserving of Federal support. As a first step, the new molecular tools should be applied to gain insight into the basic molecular and cellular processes by which marine organisms adapt to extreme environments. Examples of basic research that may lead to commercial applications are gene sequencing projects and the ongoing study of special glycoproteins that inhibit ice crystal growth in the tissues of Antarctic fish. These substances, along with gene sequence information for key marine species, may prove useful in industrial and medical preservation processes.

Second, a thorough understanding of marine ecological systems must be developed in order to specify the "normal" baseline level of function and to monitor and predict potential changes and perturbations of systems due to physical, chemical, or biological impacts. The development of predictive models for analysing potential global climate changes depends on the acquisition of fundamental information on molecular regulatory mechanisms of photosynthesis in the oceans.

Third, research on marine ecology can be conducted to benefit fisheries management. The tools of biotechnology can be used to determine the effect of natural and anthropogenic perturbations on the size of commercial stocks, to delineate fishery stocks and management units of living resources, to determine predator-prey relationships, and to restore habitats essential to robust fisheries.

Defining the Impact of the Seas on the Global Environment

Powerful tools are being developed to elucidate the many biogeochemical cycles that determine the fate of all the life-supporting elements on Earth. Scientists are beginning to understand and manipulate the molecular genetics and biology of esoteric metabolic pathways associated with the carbon, sulfur, phosphorous, iron, and other biogeochemical cycles. For example, based on the hypothesis that iron controls photosynthesis in the oceans, immunological probes were used to show that addition of iron to open ocean water off the Galapagos Islands significantly increased energy production.(23)

Marine biotechnology will be useful in assessing the role of the oceans in affecting climate change and the global carbon cycle. Molecular techniques can facilitate and enhance the measurements of CO_2 concentrations and total CO_2 inventories being developed for global ocean models of carbon cycling. There is compelling evidence that the exchange of dissolved and particulate materials between the continental shelf and its boundaries is a significant factor affecting the flux of CO_2 and biogenic elements within the global ocean. Several Federal agencies plan to collaborate on related research, including marine biotechnology applications.

Fish Health Management: Overview

When working with fish, disease prevention is always more rewarding than treatment. Once fish are sick, accurately identifying all problems present can be difficult, and treatment must be administered early in the course of an epizootic to be effective. In most cases, a comprehensive program of fish health management should be based on the principles of water quality, nutrition, sanitation, and quarantine. Water quality and nutrition are discussed below. Sanitation includes maintenance of a clean environment with minimal accumulation of organic debris, proper disinfection of nets and equipment, and thorough disinfection of fish-holding units between groups of fish. While these practices are more applicable to hatchery management and intensive, indoor systems than to pond or sea-cage culture, efforts to maintain as clean an environment as possible within the constraints of the operation will help minimize disease outbreaks. New fish should be quarantined for at least 3 wk, with 30-60 days preferred for valuable pets or zoologic specimens. Producers may be constrained by production practices; however, they may be forced to address quarantine concerns as regulations are enforced in an effort to prevent the introduction of viral diseases to aquaculture facilities.

Aquatic medicine has emerged as a recognised specialty within the practice of zoological medicine. Fish medicine, an important component of the aquatic specialty, is evolving with distinct subspecialties of aquaculture or production medicine, and pet fish medicine that focuses on individual animals. Although reference will be made to aquaculture medicine, the emphasis of this chapter is on pet fish medicine. In addition to aquaculture and pet fish practice, aquatic species are increasingly important as laboratory models for biomedical research and toxicology studies. Maintenance and husbandry of aquatic species is becoming a component of laboratory animal practice.

Many fish sold through the pet trade are raised on farms in south Florida. Live bearers (ie, guppies, platties, and swordtails) are pond raised and harvested by trapping. Most other fish are categorized as egg layers, which are often hatched indoors and then moved into fertilized ponds for grow-out. At the end of the production cycle, which is usually 3-6 mo, ponds are drained, and fish are harvested with a seine net and prepared for shipment.

A large percentage of fish sold through the pet trade is imported. Some of these fish are wild caught, while others are raised on small production facilities, many of which are located in southeast Asia. Virtually all marine fish, except for clown fish (Amphiprion spp), are wild caught. The pet trade is a global industry, and fish may be moved through several dealers before reaching a retail outlet.

Koi and fancy goldfish for ornamental garden ponds have grown in popularity. Many are imported from Japan (koi) or China (fancy goldfish). Imported, show quality fish sell for hundreds to thousands of dollars each. Koi are very hardy and quite tolerant of medical and surgical procedures. Owners value individual animals and are often eager to invest in veterinary care for these pets. Clinical management of individual pet fish, exhibit animals, and valuable broodstock has changed dramatically in recent years. Advances include use of nonlethal methods for diagnosing disease and more sophisticated treatment options. Radiology and ultrasound are particularly well suited for disease diagnosis in aquatic species. Development of blood culture techniques to accurately identify bacterial agents and run sensitivity tests prior to the start of antibiotic therapy has been particularly useful in decreasing the need to euthanize, or surgically biopsy, an animal to achieve an accurate diagnosis. Surgical advances, including use of exploratory laparotomy and swim bladder repairs, have salvaged animals that previously would have been euthanized.

Physiology

Fish are poikilothermic, and all physiologic processes are greatly influenced by water temperature. In freshwater, the internal tissues of fish are hyperosmotic, whereas in saltwater they are hypo-osmotic. Surface injuries to the skin make osmoregulation more difficult and may be of serious consequence due to loss of fluid balance and circulatory collapse.

Fish lack organised lymph nodes and Kuppfer's cells. Phagocytic tissue is located in the hematopoietic tissue of the spleen and kidney and often in the atrium of the 2-chambered heart. The structure of the fish kidney varies with the species; generally it is divided into an anterior "head" kidney and a posterior "caudal" kidney, located retroperitoneally, ventral to the vertebral column. Hematopoietic, renal, and endocrine tissues are found in the kidney. Divalent ions are excreted principally via the kidney, and monovalent ions and nitrogenous excretions via the gills. Accordingly, lesions of the kidney and gills may seriously interfere with respiration, excretion, and fluid balance.

The swim bladder in bony fish, which originates as an appendage of the foregut, regulates body buoyancy and may also be used for sound production. Gas is either secreted by or absorbed into the swim bladder to maintain buoyancy or specific gravity and balance. A sensory lateral line system along the sides of the body and head receives stimuli from the aquatic environment and mediates adaptive responses through the CNS.

A humoral antibody system occurs in all fish but varies considerably between classes. Although antibody production often is temperature dependent, specific serum antibodies can be demonstrated. B lymphocytes, found in the spleen and liver, are responsible for production of immunoglobulins found in the serum and tissue fluids of fish. However, fish lack the potent immunoglobulins similar to IgG of other animals. Fish do increase production of IgM, similar to other vertebrates, when responding immunologically to many infectious agents. Fish depend on increases in environmental temperature for efficient antibody production during infections (or after vaccinations), when most pathogens are replicating at a more rapid rate. The optimal temperature for antibody production varies with the species of fish (warmwater or coldwater). Extreme increases in environmental temperature (above that of the natural habitat) inhibit antibody production. T lymphocytes of fish, like those of other vertebrates, are responsible for cell-mediated immunity. Immunity is not as age dependent in fish as it is in other animals; young fish are usually immunocompetent and can be vaccinated successfully. Antibodies are found in the mucus of the fish skin and GI tract.

While anamnestic immunologic responses have been documented in fish, the duration of acquired immunity appears limited. Immunity lasts longer with individual parenteral administration of antigens than with mass bath methods. Although vaccination of fish against specific diseases has been economically important in preventing losses, there is a need for improved methodology.

Chapter 4

Necropsy and Diagnostic Techniques

While the same principles are used in necropsy of fish as in other animals, greater emphasis is placed on an accurate history, premortem signs, fresh necropsy material, and direct microscopic examination of fresh tissue smears and squash preparations. Fish decompose quickly, and many saprophytic microorganisms reproduce rapidly in the decaying tissues, which complicates isolation of pathogens unless samples are collected immediately after death.

A general fish necropsy may include blood collection (premortem); biopsy of gill, skin, and fin tissues; bacterial or viral culture of internal organs; and histology. A diagnostic facility that is familiar with fish necropsy protocols and aquatic microbiology should be used. Whenever possible, fish should be submitted alive. If the fish has just died, the eyes should be clear and the gills normal in colouration and texture. Freshly dead fish can be wrapped in a moist paper towel and submitted on ice. A water sample should also be submitted with the fish.

Fresh tissue samples of gill filaments, skin mucus, and fins should be collected, prepared as a wet mount, and examined under a light microscope at 100× and 400×. Fresh water should be used to prepare wet mounts of external tissues from freshwater fish, and saltwater should be used to prepare wet mounts from marine fish. If uncertain, use water from the tank or from the submitted water sample. Tissue should be examined for morphology and for the presence of parasites, fungi, or bacteria. Ensuring that the salinity used to prepare mounts is similar to the salinity present in the environment should allow organisms to remain viable long enough to allow identification. Microscopic examination of internal organs is also recommended if the fish has been euthanized. Unstained sections of stomach and intestine should be examined for the presence of parasites. Unstained sections of spleen, kidney, and liver should be examined for the presence of parasites, granulomas, or other anomalies.

Blood can be collected from the caudal vein of fish larger than 25-100 g, depending on species, and hematologic parameters measured. Use of hematology and serum chemistry is limited because normal values are not readily available; however, the information may still be clinically useful. Serology may be diagnostic in certain cases (eg, heavy metal toxicity). Whole blood (1-2 drops) can be incubated in brain heart infusion broth at room temperature on an electric rotator. If cloudiness indicative of bacterial growth develops, a loop of the blood-broth mix can be used to attempt primary isolation of a systemic bacterial pathogen.

If the fish is to be sacrificed, it should be euthanized and opened under sterile conditions. Bacterial cultures should be taken from the posterior kidney and streaked onto blood agar. Although blood agar supplemented with salt is helpful for marine fish, it is not necessary if an enriched blood agar is used. Ordal's or similar cytophage media should be available for isolation of myxobacteria (slime bacteria). Sabouraud's is an excellent all-purpose media for the isolation of fungal agents. Lowenstein's media is recommended for isolation of Mycobacteria spp-Mueller-Hinton is the media of choice for sensitivity testing of most common bacteria isolated from fish. If abscesses or other obvious anomalies are visible, those sites should also be cultured. As a general rule, bacterial or fungal cultures taken from fish tissue should be incubated at room temperature (25°C). Some agents of concern will not grow at all at 37°C, the standard temperature for incubation of cultures taken from mammals. An acid-fast stain should be available to rapidly rule out mycobacteriosis when granulomatous or other suspect material is seen. If fish are seen spinning before death, or showing other behavioural indications of neurologic disease, brain cultures are indicated. With the exception of lymphocystis and carp pox, viral diseases of pet fish have been of minimal concern to clinicians. This is changing as new diseases are being recognised. Most viral diseases of pet fish are identified using electron microscopy.

Spring viremia of carp, a disease listed as notifiable by the OIE, was reported in koi in the USA for the first time in 2002. The disease should be considered of regulatory concern in the USA. Veterinarians working with coldwater species, particularly salmonids, should be familiar with guidelines of the Fish Health Section of the American Fisheries Society for identifying and reporting viral diseases. Infectious salmon anemia, a viral disease first reported in the USA (Maine) in 2000, is of particular concern.

Therapeutic Considerations

Therapeutic options in fish are limited. FDA-approved drugs commercially available for use in food fish are listed in *Table: FDA-approved Drugs for Aquaculture Use in the USA (2002).* In addition, the FDA has listed several compounds as being of "low regulatory concern" (*Table: Drugs Designated as Low Regulatory Priority for Aquaculture by the FDA*). These compounds, though not fully approved, are considered innocuous enough for use in food fish. Of these, salt is the most important. A few compounds, including copper sulfate and potassium permanganate, are not FDA-approved, but are used in aquaculture under the provision of "moderate regulatory concern." Finally, there are several non-FDA approved compounds that are used in pet fish practice under controlled conditions. These have no legal status at present and have no place in food animal practice. In addition to being aware of FDA concerns, fish practitioners should be familiar with state environmental regulations. Federal and state environmental regulations are of greatest concern when treating outdoor ponds.

FDA-Approved Drugs: FDA-approved drugs for use in aquaculture in the USA include 2 antibiotics, 1 parasiticide, 1 anesthetic, and 1 spawning hormone. In many cases, therapeutic management of fish other than catfish or salmonids requires extra-label use of drugs.

Oxytetracycline is approved for use in Pacific salmon (for marking bony tissue), salmonids, catfish, and lobsters. It is widely available and has a broad spectrum of activity against gram-negative bacteria. Clinical evidence suggests that it is quite effective against myxobacteria, which do not grow on Müller-Hinton agar, making sensitivity testing difficult. Its main disadvantage is that it is only available in sinking feeds, which may make it difficult to determine whether it has been eaten, especially when treating fish in ponds. Because oxytetracycline has been used extensively for several decades, significant bacterial resistance has developed. Reliance on bacterial sensitivity test results is recommended.

A potentiated sulfonamide is approved for use in salmonids with furunculosis (Aeromonas salmonicida) and in channel catfish infected with Edwardsiella ictaluri . It should be fed at a dosage of 50 mg/kg for 5 days. The drug binds to the skin, and salmonid products are generally sold with intact skin, so longer withdrawal times (42 days) are required. This drug is available in floating feed, which makes it easier to determine whether it has been eaten. Clinical evidence suggests that it is not always effective against myxobacteria; therefore, it is not recommended as the drug of choice if columnaris is an important component of an epizootic.

Formalin is FDA approved for use in finfish and penaeid (saltwater) shrimp. Parasite-S is labelled for all finfish and penaeid shrimp; 2 other brands, Formalin-F and Paracide-F are labelled for select finfish species (table 7, *Table: FDA-approved Drugs for Aquaculture Use in the USA (2002)*). Methanol may be added to formalin products as a preservative. Formalin eliminates protozoan parasites and monogenean trematodes from the external surface of fish. It can be used as a prolonged bath at concentrations of 15-25 mg/L. The lower concentration is recommended for pond use because formalin removes dissolved oxygen from the water. Vigorous aeration during formalin treatment is essential. A concentration of 25 mg/L is equal to 2 drops/gal. (useful for delivering formalin to aquarium fish). When treating at d”25 mg/L, a water change is not necessary following chemical administration. At this concentration, formalin has minimal impact on biofiltration;

however, if ammonia is tested using Nessler's reagent, a very high reading may be observed for several days. This is an artifact caused by the interaction of the 2 compounds. Short-term baths with formalin can be provided at concentrations up to 250 mg/L for 30-60 min. At water temperatures >77°F (25°C), the concentration should be decreased to ~170 mg/L. Fish should never be left unattended during treatment and if adverse reaction to the chemical becomes apparent, the fish should be immediately placed in clean water. If formalin is allowed to chill to <45°F, a white precipitate, paraformaldehyde, will form. Because paraformaldehyde is highly toxic to fish, formalin should never be used if a precipitate or cloudiness is observed. Formalin is carcinogenic and potentially toxic to workers; material safety data sheets should be on hand in businesses where the chemical is used, and employees should be informed of appropriate safety precautions.

Tricaine methanesulfonate (MS-222) is FDA approved for use as a sedative and anesthetic in food fish and is often used to sedate broodstock for handling and injection of hormones for spawning. It is also useful for pet fish and is effective for sedation, surgical anesthesia, and euthanasia. Sedation can generally be achieved with concentrations between 50-100 mg/L, although species-specific sensitivities should be expected. Induction for most species may be near 125 mg/L; however, when working with unfamiliar species it is best to start at a lower concentration (ie, 50 mg/L) and increase the concentration until the desired effect is achieved. Because MS-222 is an acid, the chemical should be buffered (2 parts sodium bicarbonate by weight to 1 part MS-222). Following induction, the concentration may be decreased to 50-100 mg/L to maintain the desired depth of anesthesia. Respiration should be monitored; if opercular movement ceases, fish should be immediately moved to clean water. MS-222 can also be used to euthanize fish at concentrations of 1,000-10,000 mg/L. For small animals, a squirt bottle with a stock solution of 10,000 mg/L can be used to quickly apply a lethal dose of chemical to the gills. The compound will not remain stable for more than a few weeks when used in this manner. As it degrades, the colour of the solution will change from clear to brownish. MS-222 is light sensitive and should be kept in a brown bottle when stored as a solution.

Chorionic gonadotropin is FDA approved as a spawning aid for finfish. Veterinarians may work as part of a team for fish hatcheries and may be asked to assist in obtaining spawning hormones.

Salt: Salt can be used for many purposes, including destruction of single-celled protozoans and management of osmoregulation. Seawater is 3% salt, which is 30,000 ppm. By increasing or decreasing the amount of salt to which a freshwater fish or marine fish, respectively, is exposed, osmoregulatory stress can be minimized and many parasites eliminated. For freshwater fish, a 3% dip is an effective ectoparasiticide and is strongly recommended when moving fish. However, tolerance varies by species. Most freshwater fish will tolerate 3% salt for 30 sec up to several minutes, after which they show signs of stress, commonly manifest by rolling on their side. Recovery is rapid if fish are promptly removed from the salt solution. The use of salt is a quick, effective, inexpensive, and readily available means of minimizing the introduction of protozoans into a system with new fish.

A solution of 0.5-1.0% salt is recommended for transportation of freshwater fish, and most species will tolerate this concentration for several hours or days. A solution of 0.02-0.2% (200-2,000 ppm) salt can be added to freshwater recirculating systems as a continuous treatment to minimize parasitic protozoa in the system. Salt is less caustic than other parasiticides and seems to optimize healing of epithelial surfaces. Unfortunately, it is not practical for use in ponds (other than for control of nitrite toxicity) because of the massive quantities that would be required to achieve a nominal level of salinity. Addition of salt may be practical in small ornamental ponds of a few thousand gallons. Less information is available on lowering salinity for marine fish, but it is a technique that should not be overlooked.

Non-FDA-Approved Compounds: Copper sulfate ($CuSO_4$) is not approved by the FDA; however, a number of compounds containing $CuSO_4$ have been approved by the US Environmental Protection Agency (EPA) as algicides for use in aquatic sites. $CuSO_4$ is currently designated as "of moderate regulatory concern" and is used in food fish practice; however, practitioners must keep themselves informed of possible changes in the status of this chemical. $CuSO_4$ has been used for many years as a parasiticide and is particularly useful in large production ponds because of its relatively low cost. Copper is highly toxic to fish, and safe use depends on its interaction with carbonate salts in water. In freshwater systems, the concentration of $CuSO_4$ applied should be based on the total alkalinity (TA) of the water. If TA is <50 mg/L, copper cannot be used safely without performing a bioassay. If TA is 50-250 mg/L, a safe concentration of $CuSO_4$ can be determined by dividing the TA by 100. For example, if TA = 100 mg/L, a safe concentration of $CuSO_4$ would be 1 mg/L. If TA is >250 mg/L, the concentration of $CuSO_4$ should not exceed 2.5 mg/L. Other concerns when treating a pond with $CuSO_4$ (in addition to its direct toxicity to fish) relate to its algicidal activity. Rapid death of an algal bloom can precipitate a catastrophic oxygen depletion. Use of $CuSO_4$ in ponds not equipped with supplemental aeration is risky. Use of $CuSO_4$ is hazardous if a pond has a heavy algal bloom (secchi disc d"18 in.) or if the water is already deficient in oxygen due to other factors, (eg, cloudy weather or high water temperature).

$CuSO_4$ is efficacious against most protozoal parasites, is economical, and despite these concerns, may be an excellent choice when multiple treatments are required (eg, in an epizootic of Ichthyophthirius multifiliis). In saltwater systems, copper is sometimes applied in a chelated form because it stays in concentration longer. Chelated compounds may be difficult to use safely and require careful monitoring. $CuSO_4$ can be used to treat marine fish, but the concentration of active copper must be closely monitored (test kits are available) and should be maintained at 0.2 mg/L for up to 3 wk. Safe and effective use of copper in marine systems requires that Cu concentrations be tested at least once a day. Copper is extremely toxic to invertebrates, so these must be removed before the water is treated. Copper is also toxic to plants and should not be used in ornamental ponds that have been stocked with valuable plants. Finally, copper will impact bacteria in biofilters and a transient increase in ammonia should be expected for several days following treatment. Monitoring ammonia until measurable concentrations subside is recommended.

Potassium permanganate ($KMnO_4$) is not approved by the FDA but is also in the group designated "of moderate regulatory concern." $KMnO_4$ is used as an external parasiticide, fungicide, and bactericide. It is a strong oxidizing agent and "burns" organic material off the external surface of the fish. Overuse, particularly multiple uses within a short period of time, will kill fish. Use of $KMnO_4$ no more than once a week seems safe for many fish. The concentration of $KMnO_4$ used varies with the permanganate demand of the water.

Permanganate demand is greater in water with a high organic load than in water with little organic matter. To determine the permanganate demand, a bioassay can be performed; the water to be treated is placed in small containers and $KMnO_4$ is added in incremental concentrations of 2 mg/L. The correct concentration for therapeutic use will be the lowest concentration that maintains a pink colour for at least 8 hr. A practical method is to apply $KMnO_4$ at 2 mg/L in the morning—if the colour changes from pink to brown or clear in <8 hr, the treatment should be repeated. If the concentration of $KMnO_4$ required to maintain a pink colour for at least 4 hr is >6 mg/L, then the organic load is excessive, and sanitation practices should be evaluated. In large production ponds, little can be done to decrease the accumulation of organic material in ponds other than draining the pond, drying the bottom, and discing it. This is not done very often, perhaps once in 10-15 yr. In smaller systems (<0.1 acre), mechanisms may be in place to facilitate cleaning and removal of debris. $KMnO_4$ has little impact on biofilters when applied at 2 mg/L or less.

Hydrogen peroxide (H_2O_2, 3%) is currently used to control protozoan parasites in both food and ornamental species as an alternative to $CuSO_4$ and $KMnO_4$. H_2O_2 is categorized as "low regulatory priority" by the FDA and is used in the salmonid and hybrid striped bass industries. Food grade H_2O_2 is 35% active and is the best product for aquaculture use. The primary use of H_2O_2 is the control of sea lice and aquatic fungi on fish or fish eggs. It may also help control bacterial gill disease and columnaris. Dosages are variable and some fish may not tolerate this chemical. It is applied as a short-term bath, often for 15-30 min intervals. A dosage of 250 mg/L would be 2.7 mL of the

35% solution/gal of water. Practitioners unfamiliar with use of this compound should experiment with it in a bioassay before applying it to a valuable group of fish. Efficacy can be monitored by evaluation of gill and skin biopsy material before and after treatment.

Erythromycin is not FDA approved but has been used in management of bacterial kidney disease of salmonids and streptococcal infections in food and nonfood species. It can be incorporated into fish food at a dosage of 100 mg/kg body wt and fed for 14 days. FDA permission is required for use in food animals.

Another group of compounds have been designated by FDA as "high regulatory priority," ie, their use is likely to result in enforcement action by the FDA. The most important of these compounds are chloramphenicol, the nitrofurans, and malachite green. These compounds should never be used in food animals for any reason, and their use in nonfood species is discouraged.

Pet Fish Practice: Some drugs are used in pet fish practice that are not appropriate for aquaculture use, including both antibiotics and parasiticides. None of these compounds are approved for the uses described, and safety and efficacy data are sparse. Nonetheless, these treatments are considered somewhat routine in the practice of ornamental fish medicine.

Kanamycin has been used with some success to treat bacterial diseases of ornamental fish, including koi. It can be administered orally at 20 mg/kg, by injection at 20 mg/kg, or in a bath at a concentration of 750 mg/L for 2 hr. Anorectic fish can be medicated with a bath treatment or by injection, repeated daily, until fish begin to eat, at which time the drug can be incorporated into the feed to complete the treatment period. Treatment should be continued for 7 days beyond the alleviation of clinical signs. Aminoglycosides are toxic to fish and should be used with caution. Severe kidney lesions have been reported in goldfish treated with gentamicin. Toxicity may be exacerbated by a high ammonia concentration in the water.

Organophosphates have been used in nonfood fish practice for decades to control monogenea, crustaceans, and leeches. Historically, there was an approved compound, Masoten (used at a concentration of 0.25 mg/L active ingredient), for use in ponds stocked with nonfood fish, however the label expired and was not renewed. Use of organophosphates in aquaria is still practiced, and the dosage is often increased slightly in marine systems due to the higher pH. Use of organophosphates in outdoor ponds is not generally recommended because of legal and environmental concerns.

Metronidazole is used to control flagellated protozoans and can be delivered in a medicated food or as a bath when fish are anorectic. A concentration of ~7 mg/L (~250 mg metronidazole dissolved in 10 gal. of water) can be administered daily for 5 days. A daily water change a few hours after treatment is recommended. Metronidazole can be administered at 50 mg/kg PO, for 5 days. Anecdotal information suggests that excessive treatment (10 times the recommended dosage for 30 days) with metronidazole may be associated with reproductive failure in some fish. Fenbendazole has been used to control intestinal helminths in fish. A dosage of 25 mg/kg, delivered in food for 3-5 days, has been commonly recommended, but this regimen has not been evaluated in controlled trials.

Praziquantel has also been used successfully in fish to control intestinal cestodes as well as monogenean trematodes on the gills and skin. The most common use of praziquantel is as a prolonged bath in large marine aquaria for control of monogenean trematodes, particularly Neobenedenia spp . It is applied at a concentration of 2 mg/L, and limited work indicates that the compound remains active for several weeks. Praziquantel can also be administered PO at a dosage of 35-125 mg/kg for up to 3 days or as a short-term bath treatment at a concentration of 10 mg/L for 3 hr.

Chloroquine has been used to control Amyloodinium sp in ornamental marine fish. It is applied as a prolonged bath at concentrations of 2 mg/L. Efficacy in recirculating systems seems to be very good; however, there are essentially no data on treatment intervals, effects on biofilters, or other basic husbandry data.

Nutritional Diseases

Nutritional management of fish is highly variable, depending on species and system. For established foodfish industries, high-quality diets are readily available. For emerging industries and ornamental fish, particularly marine species, nutrition is one of the greatest constraints to development of commercial culture. Fish generally require a high-protein diet, with a significant percentage of protein provided by fish meal. Most fish do not synthesize ascorbic acid; therefore, adequate supplementation is necessary. Ascorbic acid should be provided in a stabilized form, often referred to as "stay C." Proper storage of feeds in a cool, dry place is necessary to minimize nutritional degradation.

Classic ascorbic acid deficiency has been referred to as "broken back disease" by fish farmers because of the collapse of the vertebral column that occurs secondary to inadequate collagen synthesis. Less dramatic signs of deficiency include impaired wound healing and immune function. Of the B vitamins, folic acid deficiency has been associated with poor growth and, in channel catfish, with anemia. Pantothenic acid deficiency has caused gill disease. Acute thiamine deficiency has been associated with neurologic signs, including convulsions and death; chronic deficiency has resulted in loss of equilibrium, edema, and poor growth.

Riboflavin deficiency has been associated with vascularization of the cornea, hyperpigm-entation, clouding, and hemorrhage of the eyes. Niacin, biotin, and pyridoxine deficiencies have been associated with neurologic abnormalities, including spasms and convulsions. Choline and inositol deficiency have been linked to poor growth. Vitamin A deficiency has been associated with poor growth and retinal atrophy. Vitamin E deficiency has been related to myopathy, including muscular deformities. Skeletal muscle abnormalities have been associated with selenium deficiency and rancid feeds. Rancid feeds have also been linked to steatitis. The nutritional requirements for trout, salmon, catfish, warmwater fishes, and aquarium fishes have been published by the National Academy of Sciences.

Feeding methods and rates vary with species, age, system, and water temperature. A maintenance diet is generally considered to be 1-2% of body wt/day, whereas a diet for growth may be 3-5% of body wt/day. In many systems, feeding behaviour is an important indicator of health. Vigorous feeding is desirable, and a sudden decrease in appetite should be cause for alarm. To minimize size variation within a population, feed should be distributed over as broad an area as possible. Smaller, more frequent feedings are preferable to one large feeding, especially for young fish.

Parasitic Diseases: Overview

All of the major groups of animal parasites are found in fish, and apparently healthy wild fish often carry heavy parasite burdens. Parasites with direct life cycles can be important pathogens of cultured fish; parasites with indirect life cycles frequently use fish as intermediate hosts. Knowledge of specific fish hosts greatly facilitates identification of parasites with marked host and tissue specificity, while others are recognised because of their common occurrence and lack of host specificity. Examination of fresh smears that contain living parasites is often diagnostic.

The most common parasites of fish are protozoa (Table: Protozoan Parasites of Fish). These include species found on external surfaces and species found in specific organs. Most protozoa have direct life cycles, but the myxosporidia require an invertebrate intermediate host.

Protozoa Infecting Gills and Skin

Ciliated protozoa are among the most common external parasites of fish. Most ciliates have a simple life cycle and divide by binary fission. Ciliates can be motile, attached, or found within the epithelium. The most well-known organism in the latter group is Ichthyophthirius multifiliis , which has a more complex life cycle than the other ciliates.

The infection caused by I multifiliis is referred to as "ich" or "white spot disease," although the latter name is misleading because fish can die from an infection

without ever developing the classic, white blister-like lesions. This parasite is an obligate pathogen and cannot survive without the presence of living fish. It is readily identified using a light microscope at magnifications of 40× or 100×. It is large (0.50-l mm), round, and completely covered with cilia. It has a characteristic horseshoe-shaped macronucleus. Its characteristic movement varies from constantly rotating to ameboid-like. The immature stages of the parasite (tomites) must find a host within a specified time, which varies with water temperature from several days (warm water) to weeks (cold water), or they will die. With large, complex systems, removing fish and leaving the system fallow is an acceptable means of preventing reinfection.

The tomites penetrate mucus of the epidermis and gills, where they feed on epithelial and blood cells extracted from the superficial capillaries. Once encysted under the epithelium, I multifiliis is refractory to chemical treatment. After reaching maturity, it leaves the host and sinks to the bottom sediment, where it divides within a cyst-like structure. While encysted, it is also refractory to chemical treatment, but cysts may be removed by aggressive cleaning and removal of debris from gravel substrates.

The organism multiplies within the cyst, producing 500-1,200 tomites from each adult parasite. The speed of the life cycle is determined by water temperature. The life cycle is completed in 4 days at water temperatures of 24-26°C but takes 35-40 days at 7°C. Multiple chemical treatments (with intervals determined by water temperature) are required for successful treatment of I multifiliis . At warm temperatures typical of home aquaria (eg, >26°C), infected fish should be treated every 2-3 days. Copper sulfate is the treatment of choice for pond fish, and either copper sulfate or formalin for fish in tanks. Potassium permanganate is not recommended because it is caustic to the gills, and repeated treatments may result in death of the fish.

Cryptocaryon irritans is a parasite of marine fish that closely resembles I multifiliis . Cryptocaryon spp infections are usually controlled by maintaining copper concentrations of 0.2 mg/L for 3 wk. Because copper is so toxic and difficult to use, formalin is an alternative for less experienced owners. A protocol similar to that recommended for freshwater fish is appropriate (ie, apply the chemical every third day if the tank is maintained at 24-26°C). If treating with formalin in marine tanks, the total duration of treatment should be extended. A minimum treatment time in marine systems is 3 wk. In a marine system, fish must be separated from valued invertebrates before beginning treatment as either copper or formalin will kill them. There are two important groups of ciliates that are motile and move on the surface of skin and gills of fish. These can be found in fish from all geographic areas and all types of aquatic systems. They include Chilodonella spp (which has a marine counterpart, Brooklynella spp) and the trichodinids, which are found on both freshwater and marine fish. Fish with chilodonelliasis typically lose condition, and copious mucous secretions may be noticed in areas where infestation is most severe. If gills are heavily infested, the fish may show signs of respiratory distress, including rapid breathing and coughing.

The gills may be visibly swollen and mucoid. If the skin is heavily infested, it may have a cloudy appearance, associated with excessive mucus, and the fish may be irritated as evidenced by flashing (scratching) and decreased appetite. Chilodonella can be easily identified from fresh biopsies of infected tissues. They are 0.5-0.7 mm, are somewhat heart-shaped with parallel bands of cilia, and move in a characteristic slow spiral.

Several genera of peritrichous ciliates have been grouped together and are collectively referred to as the trichodinids. These include Trichodina, Trichodinella, Tripartiella, and Vauchomia spp. Clinical signs associated with trichodinid infestation are similar to those of chilodonelliasis, although secretion of mucus is not usually as noticeable.

Trichodinids are easily identified from biopsies of infected gill or skin tissue. They are readily visible using a light microscope at 40-100×. Trichodinids move along the surface of infested tissue and appear as little saucers or, from a lateral view, as little bubbles. The body of the organism may be cylindrical, hemispherical, or discoid. Trichodinids are characterized by an attaching disc with a corona of denticles on the adoral sucker surface.

Tetrahymena corlissi , another ciliate, may be motile and surface-dwelling but is also occasionally found within tissue, including skeletal muscle and ocular fluids. Similar protozoa, Uronema spp , are found on marine fish. Tetrahymena spp are pear-shaped and 10-20 μm long, with longitudinal rows of cilia and inconspicuous cytostomes. External infestations of

Tetrahymena spp are not uncommon on moribund fish removed from the bottom of a tank or aquarium and are often associated with an environment rich in organic material. As long as Tetrahymena spp are restricted to the external surface of the fish, they are easily eliminated with chemical treatment and sanitation. When they become established internally, they are not treatable and can cause significant mortality. Fish with intraocular infections of Tetrahymena spp develop extreme exophthalmos. The parasite is readily identified by examining ocular fluids with a light microscope.

Ambiphyra and Apiosoma are a sessile ciliates that can be found on the skin, gills, and fins of fish. These seem to be more common in pond fish than tank-reared fish and have a predilection for organically rich environments. They are not generally found on marine fish. When examined from a lateral view, Ambiphyra is the shape of a tin can with a ciliated band around the middle and at the cytostome, which is distal to the attachment site.

Apiosoma spp are vase-shaped. Neither Ambiphyra spp nor Apiosoma spp are particularly pathogenic if present in low numbers (no more than 1-2/low-power field); however, when present in high numbers, these parasites can cause significant epithelial damage, predisposing fish to opportunistic pathogens in the environment and compromising respiration and osmoregulation. Infested fish demonstrate flashing, decreased appetite, loss of condition, and hyperplasia of infested epithelial surfaces. Severe infestation of the gills is particularly damaging. The organisms can be controlled with a single treatment of formalin, copper sulfate, potassium permanganate, or a salt dip. Excessive crowding and poor sanitation are frequently associated with heavy infestations and should be corrected.

Heteropolaria spp is a stalked, colonial ciliate that most frequently attaches to bony surfaces of fish, particularly the tips of fin rays and opercula. It is most common in freshwater gamefish, particularly centrarchids, (eg, largemouth bass, bluegill, and sunfish) and is frequently associated with development of "red sore disease." In the earliest stages of infection, bony protuberances appear slightly raised and erythematous; as the colony grows, they appear cottony.

Examination of material with a light microscope is required to differentiate Heteropolaria spp from fungal hyphae, and mixed infections are common. Further progression of the disease is typified by development of shallow ulcers on the lateral surface of the fish. A wet mount of fresh tissue from the margin of the lesion is required to differentiate between Heteropolaria spp , fungal hyphae, and columnaris bacteria.

Coinfection with the bacterium Aeromonas hydrophila is typical of red sore disease. If deaths occur, a single treatment of potassium permanganate or copper sulfate should be administered. If systemic bacterial infection is a component of the epizootic, antibiotics should be provided in medicated feed, if affected fish will accept a pelleted diet.

Flagellates: Ichthyobodo (Costia) spp are some of the most common and smallest (~15 × 5 μm) flagellated protozoan parasites of the skin and gills. They are flattened, pear-shaped organisms with 2 flagellae of unequal lengths. Ichthyobodo moves in a jerky, spiral pattern, and free-swimming organisms are fairly easy to identify in direct smear preparations. Once attached, the organism can be difficult to see and is often missed by the novice, but movement typical of a flickering flame is characteristic. Affected skin often has a steel-gray discolouration and gills may appear swollen.

Copious mucus production, sometimes referred to as "blue slime," is characteristic of costiasis. Behavioural signs of infestation include lethargy, anorexia, piping, and flashing. In acute cases, fish may appear hypoxic and swim to the surface frequently. In chronic cases, fish are weak, thin, and anorectic. The disease affects aquarium, pond, and marine fish. Ichthyobodo is readily controlled with salt, formalin, copper sulfate, or potassium permanganate baths. Because the parasite has a direct life cycle, a single treatment should be adequate. If reinfestation occurs, sanitation and quarantine practices should be evaluated.

One of the most serious health problems of captive marine fish is the parasitic dinoflagellate Amyloodinium spp . Its freshwater counterpart, Oodinium spp , is less common but can also result in high mortality. These parasites produce a disease that has been called "velvet," "rust," "gold-dust," and "coral disease" because of the brownish gold colour they impart to infected fish. The pathogenic stages of the organism are pigmented, photosynthetic, nonflagellated, nonmotile algae that attach to and invade the skin and gills during their parasitic existence. When mature, these parasites give rise to cysts that contain numerous flagellated, small, free-swimming stages that can initiate new infections.

Control of Amyloodinium is challenging, and the prognosis is guarded. Copper sulfate is the only therapeutic option for food animals in the USA, and repeated treatments are necessary to break the life cycle. In marine systems, chelated copper solutions can be used, but the concentration of free copper ions must be closely monitored and should be maintained at 0.2 mg/L for 3 wk. Copper is highly toxic to invertebrates, which must be removed from the system before treatment. When possible, environmental manipulation may be helpful.

Progress of the disease may be delayed by reducing lighting and by lowering temperature and organic load. For nonfood species, chloroquine has been used at 10 mg/L as an indefinite bath; it is apparently effective but is not approved for this use. Formalin and potassium permanganate are not effective in controlling Amyloodinium spp or Oodinium spp . There is some controversy as to the effect of altered salinity on these organisms, because some species seem to thrive in brackish waters. Therefore, adjustments in salinity (increase for freshwater fish, decrease for marine fish) during an epizootic should be carefully monitored for evidence of efficacy.

Internal Protozoan Parasites

Hexamita and Spironucleus spp are common, small (~9 μm), bilaterally symmetric, flagellated (4 pairs) protozoa most frequently found in the intestinal tract and occasionally in skin lesions or degenerating soft tissues of finfish. These genera are similar but differ slightly in the position and shape of their nuclei (2 within 1 organism). Pathogenicity of these organisms is variable and correlated with the number present. If there is a question as to whether treatment is warranted, the number of organisms present can be assessed in a wet mount of intestine. If <5 organisms per low-power field (LPF) are present, treatment is probably not necessary; if 5-15 organisms/LPF are present, treatment should be administered if fish are in poor condition; and if >15 organisms/LPF are present, treatment is strongly recommended. The only treatment available for hexamitiasis is metronidazole (use only in ornamental species), which should be given orally but can be administered as a bath if fish are anorectic.

The number of organisms present in infested freshwater angelfish increases dramatically after shipping and handling. Chronic problems have been seen in fish maintained in unsanitary or crowded conditions. Preventive treatment of ornamental cichlids is recommended before shipping, and broodstock should be evaluated periodically. The hatchability of eggs from heavily infested adult angelfish (freshwater) seems to be significantly decreased; resultant fry may be weak with poor longterm survival.

Cryptobia and Trypanosoma spp are slender, elongated (6-20 μm), actively motile, biflagellated protozoa that are easily detected in fresh blood and tissue smears of both marine and freshwater finfish. Hematogenous forms are generally described as Trypanosoma and have a well-developed undulating membrane. Trypanosomes may be transmitted by leeches and have been associated with anemia in blue-eyed plecostomus imported from South America. Cryptobia iubilans has been associated with granulomatous disease in African cichlids and discus.

Clinical disease is manifest by severe weight loss and cachexia. Clinically affected fish should be culled. Presumptive diagnosis can be made from microscopic examination of fresh tissue. Typically, granulomas will be found in the stomach, which may be visibly thickened. Acid-fast material will not be found in granulomas caused by Cryptobia . Motile flagellates may be visible using magnification of 400¥ or greater. Transmission electron micrographs are required to confirm the diagnosis of C iubilans.

Sporozoans: Coccidiosis, while common in freshwater or marine finfish, is rarely diagnosed in live fish. Many species of finfish are affected. The life cycles of many fish coccidia are unknown, and some involve >1 host to complete their development. In addition to intestinal infection, the internal organs also are commonly affected; sporulated Eimeria -like oocysts and sexual and asexual stages are commonly found in direct smears and histologic sections of the internal organs. Sulfamethazine, at 22-24 g/100 kg of fish wt/day in the feed for 50 days at 50°F (10°C), is used to treat food fish (21-day withdrawal time) in some countries. For aquarium fish, 10 ppm in the aquarium water once a week for 2-3 wk has been reported to be preventive, but safety and efficacy data are sparse.

Myxosporidians are common fish parasites. The myxosporidian spore consists of 2 valves, a suture line, and 1-4 polar capsules that contain coiled, extensible filaments and an infective central body called the sporoplasm. Evidence suggests that myxosporidia have indirect life cycles and use other aquatic organisms (eg, annelids) as intermediate hosts. Hence, myxosporidian infections are more common in, and more pathogenic for, wild fish or fish reared intensively in outdoor fish ponds. The organisms tend to be host- and tissue-specific. Accordingly, expression of the disease is related to the specific pathogen and host.

Myxosoma cerebralis, an important pathogen of young salmonids, is responsible for whirling disease, also known as "blacktail." Typically, infected fingerlings show rapid tail-chasing behaviour when startled, and the peduncle and tail may darken significantly. As infected fish age, skeletal deformity may result from damage to the cartilaginous structures, particularly the skull and vertebral column. Recovered fish remain carriers, and adults do not show signs, although skeletal deformities do not resolve. The disease can be prevented by purchasing uninfected breeding stock and maintaining them in an environment free of the intermediate hosts. A presumptive diagnosis of whirling disease is made by detection of spores from skulls of infected fish. Samples can be submitted to a fish disease laboratory, or procedures described by the American Fisheries Society can be followed. Diagnosis may be confirmed histologically or serologically. Whirling disease is of regulatory concern in some states.

Salmonid ceratomyxosis is caused by Ceratomyxa shasta, a myxosporidian endemic to specific watersheds in the Pacific northwest. The disease occurs in wild fish as well as in fish from hatcheries that use contaminated water. The most typical presentation includes hemorrhage and fibrinous inflammation in the posterior intestine, but other visceral organs and musculature can also be infected. Grossly, fish may appear emaciated, with a distended abdomen and hemorrhagic vent. A presumptive diagnosis can be made by examination of a wet mount of the posterior intestine and visualization of the kidney-bean-shaped trophozoites. The presence of the organism can be confirmed histologically. Some states consider C shasta a reportable disease.

Proliferative gill disease of catfish is caused by the myxosporidian Aurantiactinomyxon ictaluri . The organism has a complex life cycle, with the oligochete worm Dero digitata serving as the intermediate host. Channel catfish may be an aberrant host for A ictaluri , and the disease is usually associated with new ponds or previously infected ponds that have been drained and refilled.

Although proliferative gill disease can cause catastrophic mortality approaching 100%, losses may be as low as 1%. Disease occurs at water temperatures of 16-26°C, and mortality is exacerbated by poor water quality, particularly low dissolved oxygen or high levels of un-ionized ammonia. Gills of affected fish are severely swollen and bloody, resulting in the colloquial name "hamburger gill disease." A presumptive diagnosis can be made from a wet mount of infected gill tissue, in which filaments appear swollen, clubbed, and broken. Cartilaginous necrosis is strongly supportive of a diagnosis of proliferative gill disease; however, histology is required for confirmation.

Many species of myxosporidians produce nodular or cystic lesions in the skin, gills, muscle, or visceral organs of fish, depending on their host species and tissue preference. Henneguya is commonly found in white, cystic skin lesions of cultured channel catfish and aquarium fish; it is easily identified by the forked-tail appendage of the spore seen microscopically. If ponds are dried and limed heavily, infection can be eliminated, apparently by reduction of the intermediate hosts. Aquarium infection can be self-limiting in the absence of intermediate hosts. Henneguya may also be seen in wet mounts of gill tissue. Although an occasional cyst may be considered an incidental finding, severe damage has been associated with diffuse distribution of interlamellar cysts. Renal dropsy in pond-reared goldfish is caused by the myxosporidian Sphaerospora auratus . The disease is characterized by renal degeneration and ascites and is usually diagnosed by identification of spores in histologic sections of the kidney. Newly purchased pond-reared goldfish placed in aquaria may show signs of the disease, including death. No practical treatment is available.

The carp-dropsy complex is a disease of carp and goldfish characterized by dropsy and exophthalmos. It is associated with S angulata infection and may be complicated by viral infections (such as spring viremia of carp), carp swim-bladder disease, or bacterial septicemias. Deaths may be acute or occur over a 6-mo period. The response to drug treatment is generally poor. Proliferative kidney disease (PKD) is one of the most economically important diseases affecting salmonid industries of North America and Europe. Rainbow trout are particularly sensitive to the disease, although all salmonids seem susceptible. PKD is caused by an unnamed myxosporidian parasite, sometimes referred to as the PKD parasite. PKD has been reported in both captive and free-ranging salmonid populations. It occurs most commonly in the summer when water temperatures are >12°C, and the parasite primarily infects yearling and younger fish. Clinical signs include lethargy, darkening, and fluid accumulation indicated by exophthalmos, ascites, and lateral body swelling.

Infected fish are frequently anemic, resulting in gill pallor. Grossly, the posterior kidney appears gray, mottled, and significantly enlarged. Presumptive diagnosis can be based on observation of suspect organisms, 10-20 μm in diametre, in Giemsa-stained wet mounts of kidney tissue. Histologic examination of infected tissue, stained with H&E, is required for confirmation. Avoidance is the best preventive measure, although losses may be minimized by improved husbandry. Bacterial infections, particularly Aeromonas salmonicida , are common sequelae of PKD epizootics and, if uncontrolled, can result in substantially increased mortality. There is no treatment; however, fish that recover from the infection are resistant to subsequent outbreaks. Infected stocks in nonendemic areas should be depopulated, the premises sanitized, and disease-free stock obtained for replacement. Microsporidians are tiny, intracellular, spore-forming organisms with single polar filaments that are common parasites of finfish. They are host- and tissue-specific and can also infect helminth parasites of fish. The spores are extremely resistant.

Pleistophora ovariae infects ovarian tissue of golden shiners (bait fish), resulting in sterility. It is an important disease in the bait-fish industry. The organism has no intermediate host and is transmitted horizontally (through ingestion of infective spores) or vertically (through infected ova). Fertility declines as fish age, eventually resulting in sterility. Grossly, infected ovarian tissue appears marbled. The diagnosis is confirmed by examination of a wet mount of suspect tissue, revealing the presence of microsporidian spores. Although there is no treatment, the problem can be managed by discarding female broodstock when they reach 1 yr of age and replacing them annually. Although the young females remain infected, the disease is not yet sufficiently advanced to have a major impact on fertility.

Neon tetra disease is caused by Pleistophora hyphessobryconis , which infects the skeletal musculature of a number of species of aquarium fish, including tetras, angelfish, rasporas, and barbs. Infected fish may exhibit abnormal locomotion caused by muscle damage, and muscle tissue may appear marbled or necrotic at necropsy. The parasitic spores are readily visualized in wet mounts of infected tissue. There is no treatment for the infection, although removal of moribund fish helps prevent transmission by eliminating cannibalism within the population.

Helminthiasis

Helminths are common in both wild and cultured fish (Table: Helminth Parasites of Fish). Fish frequently serve as intermediate or transport hosts for larval parasites of many animals, including humans. Helminths with direct life cycles are most important in dense populations, and heavy parasite burdens are sometimes found. In general, heavy parasite burdens seem to be more common in fish originating from wild sources. Monogenean trematodes, which have direct life cycles, are common, highly pathogenic, obligatory parasites of the skin and gills. They are ~0.1-0.8 mm long and are best seen microscopically. The worms can be identified by their characteristic hold-fast organ, the haptor, which is armed with large and small hooks. Aquarium and cultured fish are subject to a rapid buildup of parasites by continuous infection and worm transfer to other fish in the tank or pond. Although many species are host-specific, the more common types seen in aquaria are less selective.

The 2 most common genera are Gyrodactylus and Dactylogyrus . Gyrodactylus gives birth to live young, which can be seen within the body of the adult worm, and frequently are skin parasites; Dactylogyrus lays eggs and is principally a parasite of the gills. Cleidodiscus is an important monogene found on the gills of channel catfish. Neobenedenia and Benedenia are important monogeneans in marine fish. Infected fish show hyperactivity and erratic swimming, often flashing above the water surface or rubbing the sides of their bodies against an object in the aquarium to dislodge the worms. Fish become pale as colours fade.

They breathe rapidly and distend their gill covers, exposing swollen, pale gills. Localized skin lesions appear with scattered hemorrhages and ulcerations. Mortality may be high. To prevent the disease, introduction of infected fish should be avoided. Formalin is often the treatment of choice for monogenean infestations. Multiple treatments at weekly intervals are recommended for Dactylogyrus because eggs may be resistant to chemical treatment.

Organophosphates have been used successfully in nonfood fish but are not approved for this use. Trichlorfon (active ingredient) as a prolonged bath (0.25 mg/L) is effective. Organophosphates break down rapidly as pH and temperature rise, therefore slight increases in concentration may be necessary in marine systems. A bioassay is recommended if practitioners are uncertain how to proceed. Use of organophosphates in ponds may be restricted by federal or state environmental regulations. Monogenes on marine fish can be removed using freshwater dips for 1-5 min, depending on the tolerance of the species; however, eggs will not be damaged or removed.

Trichlorfon can be used in marine systems, but some species are highly sensitive to it, particularly elasmobranchs. Increased ammonia levels should be anticipated after chemical application. Praziquantel (2 mg/L) has become the treatment of choice for monogenean infestations of aquarium fish, particularly marine species. The high cost of praziquantel is offset by excellent efficacy and target animal safety reported to date.

Digenean trematodes have complicated life cycles, with several larval stages that infect one or more hosts. With rare exceptions, the first intermediate host is a mollusc, without which the life cycle generally cannot be completed. A diagnosis usually can be established by gross or microscopic examinations that reveal the cercarial, metacercarial, or adult worms in any of the tissues or body cavities of the fish. Fish tend to form pigmented tissue encapsulations that encyst the parasites. Depending on the colour of the cysts in the skin, the condition is called black, white, or yellow grub disease. Heavily parasitized fish often are weak, thin, inactive, and feed poorly. Treatment is not recommended.

Pond-reared, juvenile, tropical fish may develop severe gill disease from metacercarial cysts in gill tissue. Although acute death is occasionally seen, infected fish more commonly die during harvest or shipping when they may be exposed to suboptimal dissolved oxygen concentrations. Treatment of infected fish has not been successful; however, prevention of the disease by elimination of the intermediate host, a freshwater snail, has been effective. Snails can be controlled in fish ponds by applying a copper sulfate treatment at night when they are active. A molluscicide, Bayluscid®, is available as a restricted-use pesticide in some areas (eg, Puerto Rico, Florida) for control of aquatic snails. It cannot be applied to ponds containing live fish but is extremely effective in eliminating snails if applied 1-2 wk before stocking.

Bolbophorus confusus is a digenean trematode that has recently been reported to cause mortality in channel catfish fingerlings in production ponds in Mississippi, Louisiana, and Alabama. The definitive host of B confusus is the white pelican, and the first intermediate host is the ram's horn snail (Heliosoma spp). Cercariae released from snails encyst in fish tissue, forming metacercariae in any tissue, but the majority are found in skin and skeletal muscle of the peduncle of juvenile channel catfish. Severe disease occurs when metacercariae encyst in visceral organs, particularly the posterior kidney and liver. Involvement of these organs can result in a presentation similar to enteric septicemia or channel virus disease, characterized by fluid accumulation in the abdomen and exophthalmia. Skin and muscle lesions typically result in raised bumps that are white to reddish in colour. Visceral involvement can result in high mortality (95%) of small fish. Heavy infestation of older fish may result in anorexia, lethargy, and loss of condition. Digenea in skeletal muscle can result in condemnation of affected carcasses by processing plants.

Ponds at greatest risk to B confusus are those frequented by white pelicans. Pelicans are federally protected; however, assistance for control of nuisance wildlife is available through Wildlife Services of the USDA. Snail control is an important part of an overall control strategy and requires a mix of chemical, biological and aquatic plant control strategies. Copper sulfate is effective against snails but will not penetrate when they are buried in mud or sealed into their shells.

Treatment is likely to be most effective in summer and early fall when snails are actively feeding. Nocturnal application of copper sulfate has been helpful in ornamental fish ponds, but care must be taken not to precipitate an oxygen depletion by killing plants and algae. Bayluscide® may be labelled in some states for control of aquatic snails. Chemical control will not eliminate snails, and efforts should be augmented by control of aquatic weeds. Snails climb emergent vegetation to lay eggs, so eliminating vegetation can decrease reproduction. Finally, biologic control may be attempted using black carp; however, these are an exotic species and stocking is prohibited in many geographic areas. Red-ear sunfish are also known to eat snails, but their potential impact on snail populations has not been tested. Due to the complexity of this problem, and rapid generation of new information, practitioners are urged to consult with extension and other aquaculture specialists.

Both larval and adult tapeworms are common in fish. Larval forms encyst in visceral organs and muscle, while adults usually are found in the intestinal tract. Aquatic Crustacea are the most common intermediate host for fish; accordingly, wild and cultured pond fish may be heavily infected. Diphyllobothrium latum , the broad fish tapeworm infection of humans, is acquired by eating larval tapeworms in the flesh of food fish. Aquarium fish may be purchased with heavy cestode infections but have limited exposure once in the aquarium (unless fed infected intermediate hosts). There is no safe, effective treatment for larval tapeworm infections. Corallobothrium spp are tapeworms occasionally found in the intestinal tract of channel catfish; however, their clinical significance is minimal.

Larval migrations of the bass tapeworm, Proteocephalus ambloplites , have been associated with reproductive failure in free-ranging populations of largemouth bass. Although usually an incidental finding, heavy infestations of tapeworms have been associated with mechanical obstruction of the lumen of the gut. The Asian tapeworm, Bothriocephalus acheilognathus , is occasionally seen in carp and aquarium fish. It is usually found in the anterior intestine and may be associated with enteritis and degeneration of the intestinal wall. Praziquantel is the drug of choice for treatment of cestodes in fish, but it is not approved for any aquatic use. It can be applied as a bath (2 mg/L for prolonged immersion or 10 mg/L for 3 hr) or in a medicated food (50 mg/kg, 1 time).

Acanthocephala (thorny-headed worms) are common in wild fish as both larval tissue stages and adult intestinal parasites. They are more common in salmonid and marine fish. Arthropods are the first intermediate host. Adult acanthocephala are easily recognised by their protrusible proboscis, armed with many recurrent hooks.

Nematodes are common in wild fish that are exposed to the intermediate hosts. Fish may be definitive hosts for adult nematodes, or they may act as transport or intermediate hosts for larval nematode forms (anisakids, eustrongylids, and others) that infect higher vertebrate predators, including humans. Encysted or free nematodes can be found in almost any tissue or body cavity of fish. Aquarium and cultured pond fish may be heavily infected if crustacean intermediate hosts are present.

Cyclops and Daphnia spp are common intermediate hosts for Philometra sp , a nematode that is pathogenic for guppies and other aquarium fish. These blood-red worms can be seen in the swollen abdominal cavity and protruding from the anus of affected fish (red worm disease).

Capillaria spp are commonly found in aquaria fish, particularly freshwater angelfish. Heavy infections in juvenile angelfish have been associated with poor growth rates and an inability to withstand shipping and handling. Treatment with fenbendazole (25 mg/kg for 3 days) is recommended, but efficacy has not been firmly established. Levamisole (10 mg/L) administered as a bath treatment for 3 days has also been recommended. Ivermectin is highly toxic to aquarium fish, particularly cichlids, and its use is not recommended.

Leeches are parasitic bloodsuckers of fish and also serve as vectors for blood parasites of fish (eg, Trypanosoma , Cryptobia , and haemogregarines). They can produce a debilitating anemia due to chronic blood loss and disease. Leech infestations are most common in wild fish, but aquarium and pond infestations can occur by introduction of infested fish, plants, etc.

Trichlorfon (0.25-1 ppm in aquarium water, use higher dosages at higher pH) is effective but is not approved for use in food fish, and environmental regulations may restrict its use in outdoor ponds. Multiple treatments may be required because eggs are resilient and juveniles may continue to hatch. Preventive measures include avoiding leeches (ie, effective quarantine) and depopulating infested aquarium fish. Infestations in recreational fishing ponds are often self-limiting.

Copepods

Some copepods, during specific stages of their complicated life cycle, are obligatory parasites of finfish. They lose their copepod form, including their appendages, and become rod- or sac-like structures specifically adapted for piercing, holding, feeding, and reproducing. Grossly, they appear as barb-like attachments to the skin or gills, where they feed on blood and tissue fluids. They can cause hemorrhage, anemia, and tissue destruction, as well as provide a portal of entry for other pathogens. Many different species of these parasites can be found on freshwater and marine fish. The anchor worms, Laernea spp , are commonly found in a wide variety of aquarium- and pond-reared fish, including goldfish and other cyprinids. Ergasilus spp infest the gills.

Lice (Branchiuria) are related to the parasitic copepods and have flattened bodies adapted for rapid movement over the skin surface. By means of hooks and suckers, they periodically attach for feeding by inserting the piercing mouth part (stylet) into the skin. Sea lice (Lepeophtheirus salmonis) are a significant disease problem of pen-reared salmonids which can be treated with hydrogen peroxide. Argulus spp are lice commonly found on aquarium, pond-reared, and wild fish. Trichlorfon at 0.25 ppm of aquarium water is the drug of choice for treating infested aquarium fish but is not approved for use in food fish. Infested fish should not be introduced.

Bacterial Diseases

Epidemics of bacterial diseases are common in dense populations of cultured food or aquarium fish. Predisposition to such outbreaks frequently is associated with poor water quality, organic loading of the aquatic environment, handling and transport of fish, marked temperature changes, hypoxia, and related stressful conditions. High concentrations of waterborne bacteria are normally found in ponds and aquaria. Many of these bacteria are opportunistic facultative pathogens, which are activated by an adverse environment, a debilitated host, or a primary pathogen. In sharp contrast, obligatory bacterial pathogens of finfish require the presence of fish for replication and are unable to survive alone for long in the aquatic environment.

Most bacterial pathogens of fish are aerobic gram-negative rods. Diagnosis is by isolating the organism in pure culture from infected tissues and identifying the bacterial agent.

Some of the most common bacterial infections that are associated with stressful conditions in freshwater systems are Aeromonas hydrophila, A sobria , and Pseudomonas spp . These opportunistic organisms frequently cause disease in freshwater fish, usually under warmwater conditions, although all species are susceptible. They produce very similar clinical syndromes, sometimes referred to as hemorrhagic septicemia. Nutritional deficiencies, traumatic injuries, parasitism, and sharp seasonal temperature changes appear to be predisposing factors. The acute form is characterized by signs of a septicemic infection with external reddening, and hemorrhages are found in the peritoneum, body wall, and viscera. Control is based on removal of predisposing factors. If antibiotic therapy is warranted, the drug selection should be based on sensitivity testing when possible.

Aeromonas salmonicida is a gram-negative, nonmotile, rod originally described as the cause of a septicemic disease of salmonids (furunculosis) and goldfish (ulcer disease). It is also a pathogen of many other freshwater and marine fishes and may produce high mortality. A salmonicida is a very important disease of koi and fancy goldfish, and is of particular concern in imported specimens. In the acute form, hemorrhages are found in the fins, tail, muscles, gills, and internal organs. In more chronic forms, focal areas of swelling, hemorrhage, and tissue necrosis develop in the muscles.

These lesions progress to crateriform abscesses that discharge from the skin surface (furuncles). Liquefactive necrosis occurs in the spleen and kidney. Diagnosis is made by isolating and identifying a pure culture of the organism from infected tissue. Avoidance through use of good quarantine practices, and vaccination when appropriate, is preferable to treatment. Vaccination is available for koi, but information on efficacy is limited. Successful treatment is possible, based on appropriate antibiotic therapy. Blood culture is an effective and nonlethal method for effective identification and sensitivity testing of A salmonicida isolates from valuable koi. Prevention and control of A salmonicida in fish and eggs is also of concern to the salmonid industry. Fish and fish eggs should be obtained from disease-free sources. Eyed fish eggs can be treated for 10-15 min in a solution containing iodine at 100 mg/L at a pH of 7 and a temperature of 10-15°C, although the drug is not approved for use on food-fish eggs. Eggs should be rinsed immediately after treatment. Antibiotic treatment should be based on results of sensitivity testing. Commercial vaccines are available for prevention of A salmonicida in salmonids.

Vibriosis is a potentially serious, common systemic disease of many cultured, aquarium, and wild marine and estuarine fishes; it is less common in freshwater fish. Vibrio anguillarium and other Vibrio spp are responsible for the disease, which produces systemic manifestations, including hemorrhages and ulcerations of the skin, fin, and tail, and hemorrhagic and degenerative changes of internal organs. Diagnosis requires identification of pure isolates from infected tissues. Isolation of V cholera from fish is not uncommon and should not cause alarm as long as the isolate is the non-O type. Preventive measures include minimizing stress and crowding. Coldwater vibriosis (Hitra disease), a serious problem in sea farming of salmonids, is characterized by high mortality, resistance to drug therapy, and stress mediation. The etiologic agent is V salmonicida . Because Vibrio spp are ubiquitous in marine environments, avoidance is difficult. Preventive vaccination with formalin-killed Vibrio is used in the salmonid industry. Antibiotic therapy should be based on results of sensitivity testing.

Yersiniosis (enteric redmouth disease) is a serious acute or chronic bacterial disease of intensively cultured salmonids. The etiologic agent is Yersinia ruckeri . Signs are darkening and hemorrhage of the mouth (red mouth), skin, anus, and fins. Chronic signs are associated with inappetence, exophthalmos, and swelling and degenerative changes of internal organs. Mortality rates are variable but are exacerbated by poor water quality and related stressors. Diagnosis is by isolation and identification of pure cultures of the organism obtained from the internal organs of infected fish. Fish that survive remain carriers and may cyclically shed bacteria, particularly when exposed to stressful conditions and water temperatures of 15-18°C. Depopulation of infected fish and avoidance of introduction of infected fish can be recommended, but preventive vaccination is the usual procedure in endemic areas. Yersiniosis can be treated successfully with antibiotics, which should be selected based on a sensitivity test. Therapy should be continued for at least 14 days.

Edwardsiella ictaluri causes enteric septicemia of catfish, the most important infectious disease in the channel catfish industry. Infection occurs in the spring and fall when water temperatures are 22-28°C, and mortality may be exacerbated by handling stress, chemical treatment, or poor water quality. The disease occurs in 2 forms-the enteric (or intestinal) form and the meningeal form. In the enteric form, infected fish may develop skin lesions characterized by massive petechial hemorrhage around the mouth, operculum, and eyes, or they may develop measles-like red punctate lesions along the body wall. There is a hemorrhagic enteritis, and the intestine may be hemorrhagic and fluid- or gas-filled. Liver lesions are common and may be evident as multifocal areas of necrosis, abcessation, or hemorrhage. In contrast, in the meningeal form, few external signs may be seen in infected fish.

The bacteria enter the CNS through the olfactory system, and affected fish develop severe meningitis. In fingerlings, the inflammation may be severe enough to erode the skull, resulting in the characteristic "hole-in-the-head" lesion. Fish affected with the meningeal form may demonstrate bizarre behaviour, including spinning, erratic swimming, and general disorientation.

Diagnosis is based on bacterial culture and isolation. Brain culture is indicated whenever E ictaluri is suspected. E ictaluri will grow on blood agar incubated at 25°C for 48 hr. Antibiotic therapy should be based on results of sensitivity testing. Oral vaccination is available for channel catfish fingerlings.

Edwardsiella tarda causes intestinal disease in a variety of aquatic and terrestrial organisms, including fish, reptiles, and mammals (including humans). In catfish, this bacterium causes a disease referred to as emphysematous putrefactive disease of catfish, descriptive of the characteristic malodorous, gaseous lesions.

Infection is usually limited to 5-10% of the population, and mortality is chronic. Clinically, affected fish may be unable to swim normally because of abnormal buoyancy created by gas-filled lesions in the skeletal musculature. When lesions burst, they are extremely malodorous. E tarda has been reported in freshwater and marine aquaria across a fairly wide temperature range, and in free-ranging largemouth bass. Clinical signs include significant ulceration of skin as well as systemic disease. Overall mortality rates are generally low, usually <5%. The organism is easily isolated using blood agar incubated at 25°C for 24 hr. Antibiotic treatment based on results of sensitivity testing is effective.

The order Cytophogales (Myxobacterales, slime bacteria, fish slime bacteria) includes an important group of opportunistic pathogens of fish that are common inhabitants of soil and water. The gram-negative, rod or filamentous bacteria have a distinctive gliding motion and form palisading masses on infected tissue. Skin or gill lesions have slimy or cotton-like surface exudates, which usually cover surface necrosis, ulcerations, and marginal hemorrhages. Flavobacterium columnaris , the member of this group responsible for columnaris disease (cottonmouth disease, saddleback), is most common in warm water and warmwater species of fish.

A presumptive diagnosis can be made from visualization of typical myxobacteria on wet mounts of infected skin or gill tissue. If the disease is diagnosed early in the course of infection, a treatment with potassium permanganate may be efficacious. If the disease becomes chronic, it may have become systemic, in which case treatment with oxytetracycline for 10 days is recommended. Columnaris disease can be confirmed by isolation of the organism on Ordal's or other cytophage media. Sensitivity tests are difficult to perform because F columnaria will not grow on Müller-Hinton media. Columnaris disease can be prevented by reducing organic loading and avoiding traumatic injuries. Cytophaga psychrophila causes coldwater (peduncle) disease and fin and tail rot. It most commonly infects coldwater fish but can be found in warmwater fish subjected to low temperatures. The lesions are especially common on the dorsal, posterior surface of the fish under the dorsal fin but may be found on any part of the body. Advanced cases show necrosis and ulceration of the peduncle. Both Flavobacterium and Cytophaga infections can be controlled by oxytetracycline.

Bacterial gill disease is a disease complex most frequently reported in young cultured salmonids or fish cultured under conditions of high organic loading. It has been seen occasionally in aquarium fish. It may be initiated by crowding and poor water quality, particularly high organic loads, high ammonia levels, and silt. Although bacteria in the genera Flavobacteria have been implicated as the cause, the association is still somewhat tentative. Opportunistic bacteria, including Aeromonas and Pseudomonas spp , may follow as secondary invaders of traumatized gill tissue. Signs of the disease are related to respiratory embarrassment due to impaired gill function. Gills appear swollen and mottled, with patchy areas of bacterial growth that can be confirmed by microscopic examination of direct gill smears.

Hyperplasia, adhesions, and deformity of the gill lamellae can be seen. In young fish affected with the disease, mortality is high and morbidity sustained. Prevention efforts include improving water quality and avoiding overstocking. A single treatment with potassium permanganate, followed by addition of salt to the system (0.02-0.5%), depending on species and type of system, may be beneficial in controlling losses, but sanitation is critical for longterm resolution of the problem. Antibiotic therapy may be used as needed to control secondary bacterial problems.

Bacterial kidney disease (corynebacterial kidney disease) is economically important in cultured salmonids. The cause is Renibacterium salmoninarum , an obligate intracellular parasite that is onc of the few gram-positive organisms that causes disease in fish. Clinically, infected fish appear lethargic and darkened. Typical lesions include grayish, localized, or conglomerate granulomata in the viscera, especially the kidney or body wall; exophthalmos; blindness; and emaciation. A presumptive diagnosis can be based on visualization of small gram-positive rods in kidney imprints. Definitive diagnosis requires isolation and identification of the bacteria by using a selective medium that contains cysteine and incubating at 15°C for 3-6 wk. R salmoninarum is transmitted both horizontally and vertically, and fish that survive an epizootic remain carriers. Infected female fish should be injected with erythromycin (11-20 mg/kg, IM) 14-60 days before spawning to prevent vertical transmission. Erythromycin (100 mg/kg for 10-21 days) is efficacious when administered in feed early in the course of an outbreak; however, it is not FDA approved for this use. Obtaining disease-free stock and preventing contamination by infected wild fish are the best preventive measures.

Gram positive bacteria of concern to fish culturists and aquarists include Streptococcus and related genera, Lactococcus , Enterococcus , and Vagococcus . Infections caused by these organisms are uncommon but can cause significant mortality (>50%) when they do occur. Chronic infections may continue for weeks, with only a few fish dying each day. Species known to be susceptible include salmonids, assorted marine fish (eg, mullet, sea bass), tilapia, sturgeon, and striped bass. Susceptible aquarium species include rainbow sharks, red-tailed black sharks, rosey barbs, danios, and some tetras and cichlids. In general, all fish should be considered susceptible.

A characteristic manifestation of Streptococcus infection is neurologic disease, often manifest by spinning or spiraling in the water column. Brain and kidney cultures from suspect fish should be incubated on blood agar at 25°C for 24-48 hr. Gram stains of pinpoint bacterial colonies reveal typical chains of gram-positive cocci, which allows a presumptive diagnosis. Confirmation requires definitive identification of the organism. Antibiotic therapy should be based on sensitivity testing. Erythromycin is often the drug of choice but it is not FDA approved for this use.

Mortality should cease within 48 hr of treatment with the correct antibiotic if complicating factors are not present. If fish are anorectic, therapy may be difficult because erythromycin does not absorb well when administered as a bath. Sources of infection may be environmental or include live foods, such as tubifex worms, amphibians, or previously infected fish. Future epizootics can be prevented if the source of infection is identified and eliminated.

Streptococcus iniae has been reported to cause disease in fish and mammals, including humans. In aquaculture, it is encountered most frequently in the intensive culture of tilapia and hybrid striped bass. Human cases have been reported rarely; in several cases individuals developed S iniae infection after handling raw fish, primarily tilapia, with open wounds on their hands. Most affected individuals had significant predisposing disease. The most common problem reported was cellulitis, and infections resolved following antibiotic therapy. When Streptococcus is isolated from aquatic species, a subculture of the organism should be sent to a laboratory that can determine whether or not it is S iniae . If confirmed, at-risk personnel should be restricted from handling infected fish.

Mycobacteriosis is a chronic or acute, systemic, granulomatous disease that occurs in aquarium fish and cultured food fish, particularly those reared under intensive conditions. Predisposing environmental factors include low dissolved oxygen, low pH, and high organic load, all of which are found in recirculating aquaculture systems. The causative bacteria can be any number of species of Mycobacterium , including M piscium , M marinum , and M fortuitum . These gram-positive, acid-fast, nonmotile bacteria are difficult to grow but can be isolated using Lowenstein-Jensen media following incubation at 25°C for 3-4 wk. Signs are variable and often resemble those of other diseases, including emaciation, ascites, skin ulceration and hemorrhages, exophthalmos, paleness, and skeletal deformities. On necropsy, gross lesions of viscera consist of grayish white, necrotic foci that sometimes coalesce to form tumor-like masses.

A presumptive diagnosis is based on visualization of acid-fast rods in granulomatous material from suspect lesions. Definitive diagnosis requires isolating and identifying the bacteria. Because the disease can produce skin lesions and an allergic dermatitis in humans, and because treatment does not eliminate the disease, infected fish are usually destroyed. Aquarists should be informed of potential risks if handling or cleaning contaminated fish or exhibits. An aquarium should be disinfected before other fish are added. Mycobacteria are not always killed by bleach; disinfection with alcohol or phenolic compounds is recommended.

Piscirickettsia salmonis has been described in salmonid species from Chile, Norway, Ireland, and Canada. Rickettsial-like organisms have been reported in tilapia, sea bass, and blue-eyed plecostomus. It is unclear whether organisms reported in different groups of fish are separate species. Of the salmonid species, coho salmon seem to be the most susceptible. These organisms do not appear to be a zoonotic threat as they do not seem able to survive at mammalian body temperatures. Rickettsial disease can result in acute mortality affecting up to 95% of fish with few gross signs. In tilapia, acute mortality events may be triggered by sudden drops in temperature. Chronic disease is manifest by nonspecific external lesions including anorexia, pale gills, and skin lesions. Internally, lesions are more typical, with granulomatous lesions possible throughout the viscera. The most characteristic lesions may be found in liver and kidney tissue, and have been described as gray to yellow mottled areas with ring-shaped foci. Histologically, intracellular organisms may be seen in macrophages and hematopoietic tissue in the liver, spleen, and kidney. Blood or tissue smears stained with Giemsa or acridine orange may reveal the intracellular organisms, often appearing as paired, curved gram-negative rods in macrophages or hepatocytes. Rickettsia-like organisms can be isolated using a variety of cell lines; however, confirmation of a suspect case may also be based on serology.

Transmission of rickettsial-like diseases in fish is not understood. In terrestrial species a vector is often required; however, R salmonis has been demonstrated to survive for 14 days in sea water, suggesting that horizontal transmission in the absence of vectors may be possible in aquatic species. Antibiotic control of these organisms appeared promising from in vitro work, but in vivo studies have not been as successful. It may be that intracellular organisms are sequestered from antibiotics.

Many other less commonly recognised bacterial diseases have been described in fish. These include pasteurellosis, Haemophilus piscium infection (ulcer disease), Eubacterium infection (fish meningitis), and others. Diagnosis of these diseases requires isolation and identification of the specific bacterial agent.

Mycotic Diseases

Aquatic fungi often are considered secondary tissue invaders that follow traumatic injuries, infectious agents, or environmental insults such as poor water quality. Because many fungi grow on decaying organic matter, they are especially common in the aquatic environment. Fish egg masses, which usually contain tissue debris and dead ova or embryos, are especially vulnerable. Iodophors of varying iodine concentrations are used to prevent mycotic infections of nonfood-fish eggs, which can be disinfected by a 100 ppm iodine bath for 10-15 min. This solution is toxic for hatched fish, and only eggs should be treated. Formaldehyde, up to 2,000 ppm for 15 min, can be used to treat eggs of food fish (salmonid and esocid) for the control of fungi. Chronic fungal infections of eggs may suggest improper incubation temperature or poor sanitation.

Saprolegnia infections are among the most common fungal infections of fish and fish eggs. Gross signs are grayish white, cotton-like growths on the skin, gills, eyes, or fins that may invade deeper tissues of the body. Microscopically, saprolegniasis can be recognised by making direct smears from the infected tissues and observing the nonseptate hyphal elements and mycelia. The sexual stages of the fungus can be seen only in cultures of the organism and are required for specific identification. Sabaroud's dextrose agar is acceptable for primary isolation of oomycetes, including the genera Saprolegnia . Preventive measures include removal of predisposing causes, eg, inadequate sanitation, excessive chemical treatment, or the presence of dead, infected fish and decaying organic material. If the environment is clean and skin pathogens have been eliminated, a single treatment with potassium permanganate is often adequate to control external saprolegniasis.

Ichthyophonus hoferi causes an internal fungal infection. It is occasionally seen in wild fish as well as in aged cultured and aquarium fish. The disease is generally chronic and progressive. It usually is detected on necropsy when the characteristic spherical cyst stages are seen microscopically in the smears of granulomatous lesions of the heart, liver, spleen, kidney, skin, and muscle. Lordosis and scoliosis are occasionally seen in infected rainbow trout. Culture of suspect lesions is recommended, and Sabaroud's dextrose agar is acceptable for primary isolation. Preventive measures include removing infected fish and avoiding feeding raw fish products.

Branchiomycosis is a fungal disease of gill tissue characterized by respiratory distress and gill necrosis. The causative agents are Branchiomyces sanguinis and B demigrans , which are opportunistic pathogens found in decaying organic materials in the aquatic environment. Branchiomycosis typically occurs in warm ponds (water temperature >20°C) with high organic loads. It seems endemic in eastern Europe, where it is an important problem in food fish industries; it is also occasionally reported in the USA. Affected fish show signs typical of hypoxia. Gross examination of gills reveals necrotic, mottled, and pale tissue. A presumptive diagnosis can be made by observation of characteristic aseptate, branched hyphae in gill lamellae. Branchiomycosis is prevented by avoiding predisposing conditions. Ponds that are at risk should be kept as clean as possible, and stocking rates decreased when water temperatures approach 20°C.

Many other less common mycotic infections have been reported in fish, including infections with Achyla , Aphanomyces , Dermocystidium , Ichthyosporidium , Basidiobolus , Phoma Candida, Cladosporium, Fusarium, Penicillium, Ichthyochytrium spp, and others. Fusarium is emerging as an important disease of captive marine fish, particularly elasmobranchs. Many of these organisms have questionable status as fungi, and laboratory culture of their complete life cycle is required for accurate diagnosis.

Viral Diseases

Descriptions of viral diseases of fish are rapidly expanding. Viruses are being reported in new species and interpretation of the significance of findings is also changing. The OIE lists 5 viral diseases of fish as notifiable: epizootic hematopoietic necrosis, infectious hematopoietic necrosis, spring viremia of carp, viral hemorrhagic septicemia, and Oncorhynchus masou virus. In the USA, these diseases should be reported to the USDA. Infectious salmon anemia is not listed as notifiable by the OIE in the 2002 edition of the International Aquatic Animal Health Code; however, it is considered a reportable disease in the USA. Sources for current information on regulatory requirements for aquatic species include the OIE, USDA, state veterinarian offices, state departments of agriculture, and university extension veterinary aquaculture specialists.

While viruses of homeothermic animals are cultured at uniform temperatures, fish viruses have wider, but specific, temperature tolerances in fish cell cultures at lower temperatures. Because of this relatively defined temperature range, variation in temperature may enable control, although often it merely induces latency. Because many viral diseases of fish are geographically limited, regulatory agencies and fish farms in disease-free areas consider them as exotic diseases and require certification of introduced stocks. Many produce high mortality in young fish and little or no losses in adults, which may become carriers. For these reasons, avoidance of carriers and certification of SPF replacement stocks are frequently required. Specific testing procedures are available. Vaccines are not yet commercially available for viral diseases of fish. Drugs are not effective, although antibiotics and other drugs may be used to control secondary bacterial infections that frequently follow viral diseases. Management techniques that minimize stress and crowding, biosecurity measures, and temperature manipulation hold the greatest promise for control of many piscine viral diseases.

Herpesviruses: Channel catfish virus (CCV) disease is an acute, virulent herpesvirus infection of fry and fingerling channel catfish that can cause mortality of >80% at water temperatures ?25°C in small fish (?5 cm). As fish age, mortality decreases, and clinical infection in fish >1 yr old is rare. Acute infection often includes a recent history of a stressful event such as handling or transport, low dissolved oxygen, or chemical treatment. Infected fish show signs of ascites, exophthalmos, and hemorrhages in fins. The cell line of choice for virus isolation is channel catfish ovary, followed by serum neutralization to confirm identification. Typical cytopathic effects include cell fusion, syncytia formation, and intranuclear inclusions. There is evidence for vertical transmission of CCV; consequently, survivors of an epizootic should not be used for broodstock. Although CCV can cause severe mortality when an outbreak is in progress, the annual number of cases of CCV in the catfish industry is relatively low.

Herpesvirus disease of salmonids has been subdivided into 2 types, HPV-1 and HPV-2, based on serology and DNA homology. HPV-1 has been isolated from salmonids in the western USA only, from ovarian fluid of moribund rainbow trout, from steelhead trout dying of other causes, and experimentally from chinook salmon. Signs of HPV-1 in prespawning rainbow trout included darkening and fluid accumulation, as evidenced by exophthalmos, ascites, and edema of visceral organs and skeletal muscle. A presumptive diagnosis is based on observation of the typical herpesvirus cytopathic effect, including giant cell formation and Cowdry Type A inclusion bodies, in salmonid cell lines. Confirmation requires serum neutralization. HPV-2, previously referred to as Oncorhynchus masou virus, refers to a group of salmonid herpesviruses from Japan, all of which are oncogenic except for one strain, NeTVA.

These viruses are on the list of 5 fish diseases categorized as notifiable by the OIE. HPV-2 has been isolated from masou, kokanee, coho, and chum salmon, and experimentally from rainbow trout. The virus was initially isolated in 1978 from ovarian fluid of clinically normal masou salmon that were producing fry with poor survival rates. Fish with clinical disease may be lethargic and anorectic. Darkening and petechiation of the body wall may be evident. Tumorigenic strains result in development of epitheliomas on fins or mandibles of survivors. A presumptive diagnosis is made from observation of the typical herpesvirus cytopathic effect on salmonid cell lines. Confirmation requires serum neutralization.

Herpesvirus disease of turbot is a disease of wild and cultured turbot that causes massive hypertrophy and fusion of epithelial cells of the skin and gills of young fish. Mortality is associated with heavy gill infections and poor water quality. High levels of oxygenation are essential for fish with respiratory distress. Diagnosis is by examination of skin scrapings or histologic sections for the characteristic fusion of giant cells.

A recently described herpesvirus has been blamed for serious disease in koi, with mortality rates reported to reach 100%. The disease is characterized by severe necrosis of gill epithelium. Affected fish are lethargic and show signs of respiratory distress, eg, piping and laying near inflows of oxygenated water. Grossly, copious mucus secretion may be noted on the gills and skin, and gross necrosis of gill tissue may be evident. Presumptive diagnosis of koi herpesvirus may be based on histologic evidence of epithelial necrosis of gill tissue and possible intranuclear inclusions. Transmission electron microscopy should demonstrate herpesvirus particles within inclusions. Laboratory tests to confirm the diagnosis are not currently available. Currently, koi affected with systemic herpesvirus infection should be destroyed and contaminated facilities disinfected.

Carp pox is one of the oldest recognised fish diseases. It is caused by cyprinid herpesvirus-1. Pox lesions may occur on other species of fish, and are sometimes referred to as fish pox. Lesions typically are smooth and raised and may have a milky appearance. They are benign, non-necrotizing areas of epidermal hyperplasia. Severe cases may result in development of papillomatous growths, and these may be a site of complicating bacterial infection. Generally, lesions are self-limiting and of minimal clinical significance. Carp pox can be a significant problem with koi because the aesthetic quality, and hence the market value, of the animal is severely compromised. For the serious koi enthusiast, carp pox-affected fish should be culled, preferably during quarantine. Surgical removal of pox lesions has not been rewarding.

Rhabdoviruses: Infectious hematopoietic necrosis is caused by a rhabdovirus and is endemic in salmonid (Oncorhynchus spp) populations in the Pacific Northwest and Alaska. The disease is listed as notifiable by the OIE. Disease has been reported in Atlantic, chum, chinook, sockeye, and kokanee salmon and in cutthroat, steelhead, and rainbow trout. Lake trout and Arctic char, members of the genus Salvelinus , appear resistant. Experimental infection of nonsalmonid fish demonstrated mortality of 43% and 87%, respectively, in sea bream and turbot. Infectious hematopoietic necrosis has also been isolated from leeches, parasitic copepods, and mayflies. Sporadic epizootics have been reported in several states in the USA and in Japan, Italy, France, Taiwan, Belgium, and Korea.

Most epizootics have been attributed to importation of infected eggs or fry. Acute disease in fry <2 mo may result in high mortality (>90%) with few external signs. Disease usually occurs at water temperatures of 10-12°C, although outbreaks occasionally occur at temperatures >15°C. Typical signs include lethargy with sporadic hyperexcitability, including whirling. Sick fish may be darkened with distended abdomens, exophthalmia, pale gills, and mucoid fecal casts. Important differential diagnoses include infectious pancreatic necrosis and viral hemorrhagic septicemia.

Hematopoietic tissue in the kidney and spleen are most severely affected by necrosis. Risk factors include age (fish <2 mo are most susceptible), density, and water temperature. Hauling young fish around dams in trucks may be a significant risk factor because of crowding during transit. Although most disease outbreaks have been reported in freshwater, active disease has occurred in Atlantic salmon housed in sea cages. Diagnosis is by viral isolation (from kidney and spleen of young fish and ovarian fluid of broodstock), with confirmation by serum neutralization. Rapid serologic tests have been developed and are becoming more available.

A nonlethal test involving viral isolation from mucus has been reported. The disease is transmitted horizontally through the water, and vertical transmission seems likely. There is some controversy as to whether the virus is present on the outside of the egg or within the egg, as disinfection with iodophors (100 mg/L) seems to stop >99% of vertical transmission. Asymptomatic carrier fish serve as reservoirs of infection.

Viral hemorrhagic septicemia, the most important viral disease affecting cultured rainbow trout in Europe, is caused by a rhabdovirus in the family rhabdoviridae. The disease is believed to have spread to Europe >100 years ago when rainbow trout were first imported from the USA. The disease causes marked necrosis of hematopoietic tissue in the kidney, particularly the anterior kidney, but largely spares excretory tubules in the posterior kidney.

In addition to rainbow trout, brook and lake trout (genus Salvelinus), and Atlantic salmon and brown and golden trout (genus Salmo) are susceptible. The virus also causes disease in a variety of freshwater and marine coldwater fish including pike, turbot, white fish, and sea bass. New information indicates that viral hemorrhagic septicemia is found in free-ranging marine fish in the Pacific Northwest including anadromous salmon (coho and chinook) as well as haddock and cod in the North Sea.

The disease occurs in 3 forms-acute, chronic, and nervous. Acute mortalities occur in rainbow trout fry <3 g and <30 days old. In these fish, the kidney is swollen and the anterior segment is necrotic and pale. The liver may be pale with hemorrhagic mottling, and systemic hemorrhage may be visible in the eyes, skin, skeletal muscle, and viscera. The most notable lesion is widespread hemorrhage in the liver, adipose tissue, and within skeletal muscle. Moribund fish lie on the bottom of the tank and may exhibit sporadic flashing and corkscrew swimming behaviour. As fish age, mortalities drop from 80-100% to 10-50%. The chronic form is a persistent infection characterized by the ability to isolate virus from all tissues and low red and white blood cell counts. Chronically infected fish may exhibit few visible external signs.

The nervous form of the disease has been reported primarily in cultured freshwater fish but was recently reported in marine fish. The optimal temperature for active infection is 9-12°C; the virus is unable to replicate at temperatures >15°C. The cell line of choice for virus isolation is bluegill fry (BF-2). Viral identification is confirmed by serum neutralization. Newer diagnostic tests include immunofluorescence, ELISA, and PCR. Viral hemorrhagic septicemia is a heavily regulated disease with disease-free geographic regions defined in Europe. No vaccine is commercially available.

Spring viremia of carp (SVC) is an acute, virulent, usually hemorrhagic disease of cultured carp caused by Rhabdovirus carpio. The disease is listed as notifiable by the OIE. Historically, it was reported in Europe and the former USSR; however, an outbreak in the USA was confirmed in the spring of 2002. SVC causes disease in common carp, including koi, as well as grass, bighead, silver, and crucian carp. Limited experience suggests that common goldfish may be susceptible. Clinical signs are nonspecific and may include darkening of the skin, exophthalmia, ascites, pale gills, hemorrhage, and a protruding vent with thick mucoid fecal casts. Pinpoint hemorrhage in the swim bladder is indicative of SVC, if present. Coinfection with Aeromonas or other systemic bacteria may obscure the presence of the virus and is referred to as carp-dropsy complex. The bacterial component of the infection can be controlled with antibiotics. The disease causes death in both adult and young fish. Affected fish may lie on their sides on the bottom of the tank, lose motor control, have ascites, and show petechiation of the skin, gills, and visceral mass. Clinical disease may be controlled by raising water temperature >22°C; maximum mortality can be expected at temperatures <18°C. The virus is readily isolated in common fish cell lines and identified by serum neutralization and fluorescent antibody tests.

Pike fry rhabdovirus disease is an acute hemorrhagic infection of young northern pike, thus far known only in Europe. Affected fry have pale gills, exophthalmos, and hydrocephalus. Kidney tubules show degeneration and necrosis. The causal agent is readily isolated in a variety of cell cultures. Identification is by serum neutralization.

Infectious pancreatic necrosis is an acute, systemic, contagious disease of salmonid fry and fingerlings cause by a birnavirus. The virus is the archetype of the aquatic birnaviruses, which are further subdivided into 2 serotypes, A and B, that do not crossreact using serum neutralization. The Serotype B group currently consists of only 10 isolates, all European in origin. In contrast, the A serotype contains >200 isolates that have been further subdivided into 9 serotypes, A1-A9. Morbidity and mortality occur only in young animals, usually <3 g; however, virus can be isolated from survivors for the duration of their lives resulting in a persistent carrier state. Recrudescence of disease in survivors has not been reported.

The virus is vertically and horizontally transmitted, widespread, and reported worldwide, except in Iceland and Australia. Rainbow trout are highly susceptible to disease. In the USA, striped bass and their hybrids are recognised as potential carriers. Other species affected include freshwater eels (Anguilla spp), yellowtail, turbot, sea bass, and menhaden, as well as aquatic invertebrates including molluscs and crustaceans. Brook trout are believed to be reservoirs of infection in the USA. Clinical infection is nonspecific. Diseased fish may be anorectic, ataxic, and display a corkscrew swimming pattern. Externally, fish are darkened; exophthalmia and external petechiation may be evident. Internally, petechiae may be visible on viscera; the gut is typically empty and may contain a yellow exudate. Fecal pseudocysts may be evident in the water column. Histologically, focal areas of coagulative necrosis involve acinar and islet cells of the pancreas and hematopoietic cells of the kidney. Intracytoplasmic viral inclusions may be visible in pancreatic acinar cells. Infection should be confirmed with viral isolation followed by serum neutralization. Most fish cell lines are susceptible. The virus can also be identified using fluorescent antibody, complement fixation, and ELISA techniques. There is no treatment for infected fish, but avoidance can be accomplished by purchase of SPF stocks, quarantine, and disinfection of eggs with iodophores (20-50 mg/L). Infectious pancreatic necrosis is not regulated by the USDA, but state regulations do exist in various parts of the country.

Iridoviruses: Lymphocystis disease is a unique, typically chronic, viral infection of wild or captive marine and freshwater fish. The causal agent is an icosahedral DNA virus ~300 nm in diametre, of the Iridoviridae family. Infection may be manifest by benign cauliflower-like lesions typically located on fins. The disease affects a wide range of fish and is generally considered global. Within the aquarium trade, painted glass fish are commonly infected. Presumptive diagnosis is based on the presence of enlarged fibroblasts (up to 1 mm), which are easily visualized with a light microscope. Microscopic examination typically reveals the appearance of grape-like clusters of virus-laden cells. Diagnosis is confirmed histologically. Feulgen-positive cytoplasmic inclusions and a hypertrophied nucleus are pathognomonic. The disease is usually self-limiting but is of aesthetic concern.

Viral erythrocytic necrosis is an OIE-notifiable disease. It has been reported in >20 species of marine and anadromous fish (both cultured and free ranging) and is characterized by erythrocytic degeneration. Affected species include Pacific herring, Atlantic cod, and Pacific salmonids (chum, pink, coho, and chinook), steelhead trout, and cultured eels in Taiwan. The disease is chronic, and external signs may be subtle or nonexistent. Sick fish are anemic, which may result in pale gills and internal organs. Severity of the disease is related to age and species of fish, with juveniles <1 g most severely affected. The characteristic lesion is a single eosinophilic cytoplasmic inclusion body in the circulating erythrocytes of anemic fish. The inclusions are best visualized from Giemsa-stained fresh blood smears. Transmission electron microscopy has demonstrated hexagonal virus particles, presumptively classified as an iridovirus, in the cytoplasm of cells containing inclusions. To date, the agent has not been successfully isolated. Histologically, increased hematopoietic activity may be evident in the kidney, and round cytoplasmic inclusions (0.8-4 μm) are found in circulating RBC.

Inclusions stain pink or magenta with Giemsa. Other degenerative changes may be evident in RBC, including cytoplasmic vacuolation and margination of nuclear chromatin. Hemolytic anemia with concurrent hemosiderosis and erythroblastosis has been reported in moribund Pacific herring. Multinucleated giant erythroblasts may occasionally be seen in peripheral blood, and macrophages may phagocytize abnormal erythroblasts.

A presumptive diagnosis is based on the presence of typical cytoplasmic inclusions in circulating erythrocytes of anemic fish. Confirmation requires visualization of hexagonal virus particles in cytoplasm of affected erythrocytes using transmission electron microscopy. A marine reservoir is suspected but has not been identified. Vertical transmission is suspected due to the high prevalence of infection in fry from infected broodstock.

The ranaviruses are an important group within the family Iridoviridae that affect fish. One of these, epizootic erythropoietic necrosis (EHN) is listed as a notifiable disease by OIE. It was first reported in redfin perch in Australia in the spring of 1984, but has also been shown to cause disease, albeit less severe, in rainbow trout. Similar viruses have been reported in sheatfish in Germany and black bullhead catfish in France and Italy.

EHN is endotheliotropic, producing necrotic lesions in the endothelium of blood vessels and some visceral lesions. Behavioural signs include lethargy, darkening, and erratic swimming. Mortality occurs after 4-5 days. The most consistent lesion associated with EHN is focal necrosis of hematopoietic tissue in the anterior kidney and liver. Necrotic hematopoietic cells may be visible within blood vessels. Presumptive diagnosis is based on clinical signs and isolation of the suspect agent in cell culture. Bluegill fry (BF-2) is the cell line of choice. Detection may also be accomplished using ELISA, immunofluorescence, or electron microscopy. Epizootics of EHN in redfin perch are most common in the spring and summer and almost exclusively involve juvenile fish. Survivors seem to be resistant to future infection. There is no evidence of vertical transmission of EHN, and redfin perch carriers have not been detected. An unidentified reservoir and carrier host is suspected. Fomite transmission of EHN has been demonstrated, and birds have also been shown to carry infected material.

Largemouth bass virus is a ranavirus that was isolated from moribund largemouth bass in South Carolina in 1995. It was previously isolated from largemouth bass in several Florida lakes but had not been directly associated with disease. In recent years it has been found in largemouth bass in most southeastern and many midwestern states. The disease is not well understood because the virus is commonly isolated from tissues of clinically normal fish. In the 1995 fish kill, ~1,000 fish died over a 2-3 mo period in an area that encompasses >66,000 hectares. Lesions were nonspecific and are still poorly described. Fat-head minnow (FHM) is the cell line of choicc for isolation of virus.

Several other iridoviruses have been described in ornamental fish. Two of these were initially reported as being closely related to largemouth bass virus, but more recent work indicates that the isolates from guppy and doctor fish are not as closely related as originally thought. An iridovirus has been described in freshwater angelfish (Pterophyllum scalare) showing signs of systemic disease, but the agent has not been isolated. An iridovirus has been isolated from gouramis in the genus Trichogaster using a tilapia heart cell line. This virus does not grow on FHM cells or other common cell lines used for isolation of fish viruses. The gourami virus has been associated with systemic disease and mortality of Trichogaster spp gouramis. Efforts to fulfill River's postulates were supportive though not conclusive. Clinical disease with largemouth bass virus and the gourami iridovirus seems more severe at water temperatures ?30°C, based on very limited information.

Infectious salmon anemia is an emerging disease in USA aquaculture. It is classified as a significant disease by OIE, but the USDA has listed it as a federally notifiable disease in the USA. The first report was from farmed Atlantic salmon on the west coast of Norway in 1984. Affected fish were lethargic and severely anemic (PCV <5% in moribund fish). The causative agent is an orthomyxovirus. Acute outbreaks result in high mortality. Initial signs include lethargic fish hanging around the edges of the cage. As the disease progresses, moribund animals lie on the bottom.

The most obvious external lesions are pale gills and hemorrhage in the anterior chamber of the eye. Internally, the liver appears dark and hemorrhagic, an important indication of infectious salmon anemia. Other lesions may include a fibrinous capsule around the liver, a distended stomach filled with viscous mucus, and sometimes hemorrhagic areas on the mucosa. Infected fish often have obvious ascites, and hemorrhage may be present in skeletal muscle.

Histologically, the most important lesion is multifocal, hemorrhagic hepatic necrosis, which may appear zonal; hepatocytes may be dark and swollen and necrotic areas are eosinophilic. Circulating RBC are small, and evidence of cytoplasmic vacuolation, nuclear degeneration, and cell fragmentation may be seen. Affected fish may develop lymphocytopenia and thrombocytopenia and an apparent increase in immature RBC in the peripheral circulation.

Signs of chronic infection are more subtle but may include hemorrhage in the swim bladder and skin. Diagnosis is based on clinical signs, with emphasis on anemia (PCV <10%), the gross appearance of a dark liver, and hepatic necrosis. Confirmation can be by viral isolation using the SHK-1 cell line. Virus may be visualized in endothelial cells of cardiac blood vessels using transmission electron microscopy. The agent is enveloped, slightly pleomorphic, and ~100 nm in size. Suspected cases can also be verified using an immunofluorescent antibody technique on frozen tissue. Transmission is horizontal and virus is shed in skin, mucus, feces, and urine. Sea lice (Lepeophtheirus salmonis) may be a vector; disease outbreaks seem worse when sea lice are present. Currently there is no evidence of vertical transmission. Sea trout have been proposed as a possible reservoir of infection. Protective immunity has been demonstrated in salmon that survive an outbreak. The disease is heavily regulated in Norway and now in the USA, where the USDA should be notified immediately of any suspected cases.

Miscellaneous Noninfectious Diseases

Wild fish populations are monitored for detection of toxic pollutants. Many sources indicate that the industrial and domestic production of sulfur dioxide fumes from the combustion of fossil fuels has been responsible for acid rain and the resultant high mortalities in fish in natural bodies of water. The formation of toxic aluminum compounds in acid waters also has been incriminated as a cause of fish morbidity and mortality. The discharge of toxic substances into natural waters has resulted in fish kills; some pollutants are being investigated as carcinogens because of a high incidence of tumors in fish in defined bodies of water, both marine and freshwater.

Higher incidences of tumors have been found in bottom-dwelling species and are thought to be related to carcinogens in the sediment. Ulcerative dermatitis and fin and tail rot have been described in marine fish such as flounder, salmon, sea trout, and others. Although aquatic pollutants have been incriminated, the etiology of these conditions remains obscure.

Cultured, pond-reared food fish are subject to environmental pollutants such as industrial pollutants, fertilizers, pesticides, road salt, and other runoff from surface drainage into fish ponds. Aquarium fish also have environmental problems due to accidental or intentional introduction of toxic substances, including drugs for medication, disinfectants, soaps, and aerosol sprays.

Feed intoxications of fish are commonly due to spoilage and rancidity. A dramatic food intoxication is aflatoxicosis, which results from ingestion of fish feeds contaminated with aflatoxin produced by Aspergillus flavus . Rainbow trout and other salmonids are particularly susceptible to aflatoxicosis, which results in the induction of rapidly growing hepatomas and high mortality. Other toxin-producing molds can produce similar intoxication.

Pansteatitis and swim-bladder thickening due to toxic cottonseed in fish meals incorporated in a diet low in vitamin E are reportedly associated with high mortality in cultured fish. Contaminated binders, used in making pelleted fish feeds, and salt contamination have been reported. The latter may result from errors in feed formulation or from contamination with seawater during storage. Heavy metals, plastics, oil, phenolic compounds, and other organic substances have been reported to be toxic for fish, which have become standard laboratory animals for bioassays of such toxic substances.

Neoplastic diseases similar to those found in other animals are found in fish. Their incidence frequently is higher in some geographic areas and in certain species. Some tumors are genetically mediated, such as the malignant melanoma of the gypsy-swordtail cross, and possibly the pseudobranch tumor of cod, thyroid tumors, malignant lymphosarcoma of northern pike, and fibromas or sarcomas of goldfish. The reported incidence of tumors in sharks, skates, and rays is low.

Gonadal tumors are important neoplastic disorders of koi. Typically fish present with a swollen abdomen, and depending on the severity of disease, there may be significant loss of condition. The presence of a mass can be confirmed with ultrasound. Biopsy of tissue may not offer a clear diagnosis. Laparotomy of affected fish often reveals a circumscribed mass of gonadal tissue. Fish that are not excessively debilitated are excellent surgical candidates for removal of the mass.

Colouration anomalies and yolk-sac anomalies or deformities are common in cultured fish and may be of genetic or environmental origin. For example, blue-sac disease, a condition of larval rainbow trout, is believed to be associated with unsuitable hatchery water, and pseudoalbinism in cultured flatfish is associated with excess light levels shortly after hatching.

Sunburn can occur in surface swimming fish or can be induced (even in bottom-dwelling species) by feeding photodynamic drugs such as phenothiazine, although ultraviolet light penetrates water poorly. Nephrocalcinosis and visceral granuloma are found particularly in salmonid culture, induced supposedly by a high level of carbon dioxide in the water; this produces a metabolic acidosis and urinary and tissue precipitation of calcium, around which extensive granulomata develop.

Genetic Improvement of Fishes

In contrast with livestock and plant crops where improvements in production have been based on modern breeding approaches, only a few examples of such breeding programmes exist for fish (e.g. Atlantic salmon in Norway, Nile tilapia in Asia, channel catfish in the USA). Thus, principles of genetics can be applied to the farming of marine species to increase cost effectiveness through improved breeds. The application of such technologies can be divided into two broad groups: those for short-term and those for long-term improvement.

Opportunities and Dilemmas in Molecular Aquaculture Genetics

Introduction: Transgenic vs Enhanced Selection

Aquaculture, like agriculture, is beset with problems of cost, price, and risk. Genetic research, in aquaculture as in agriculture, promises – or dangles – solutions to many of these problems "real soon now". But what kinds of solution? Investors as well as farmers soon notice that there are two, competing styles of modern genetics, which in this paper are loosely termed "transgenic" and "enhanced selection".

These styles are in vigorous competition for public attention and resources. History suggests that the first genetic style to gain a significant lead in solving an aquacultural problem is likely to drive the other style out of the game, probably forever. Inbred/hybrid corn technology effectively stopped the development of open-pollinated, pure-line varieties of corn (Kloppenburg, 1988), and the developers of fourway hybrid poultry have driven pure-line poultry breeding to commercial extinction in developed countries. Commercialization of transgenic salmon which grow twice as fast as existing strains could happen as early as this year, 2004 (Stokstad, 2002; Hoag 2003); if this were to occur classical salmon breeding programs could very well be put out of business. Transgenics with 2X or

4X current commercial growth rates are also available for species of tilapia (Rahman et al., 2001). In tilapia, however, the rate of broodstock improvement through wellconducted selection is so rapid (e.g. Bolivar and Newkirk, 2002) that step-wise transgenic technology may be unable to keep up or even gain a toe-hold. Both styles of genetics provide useful solutions to aquacultural problems but they have rather different consequences for the industry as well as for the consumer. The winning style in this technology race (more likely than not a winner-take-all race) cannot yet be predicted. This paper describes recent work which is representative of the two styles, in sufficient detail that the current level of sophistication of the science and the size of the outstanding problems can both be appreciated. The topics include possible biological limitations on transgenic growth, progress in transgenic growth and disease resistance, QTL selection, the crucial role of the major histocompatibility complex, and prevention of wasteful reproduction and pointless sexual dimorphisms.

The winner in the commercial (as opposed to technological) race between selection and transgenics will depend to a large extent on the willingness of the public to accept genetically engineered fish. Transgenic fish are looked at with strong disfavour at the present time (Hulata, 2001), but in the long term public opinion is an incalculable quantity. The two styles of genetics also have have different longterm consequences for farmers. Marker-assisted selection and genetic engineering can benefit the owners and developers of the technology and, eventually, the fish-eating public. They will also give transient economic benefits to "early adopter" farmers, but, in the long term, only those farmers who hold exclusive franchises or own part of the technology are likely to benefit from it.

The Impossibility of Standing Still

Most farmers are unwilling simply to ignore genetics, for two reasons. One is the obvious opportunity cost – lost opportunity – of forgone genetic improvement. This cost, which can be estimated from simple economic models of the genetic process, is a commonplace of the promotional literature of aquaculture genetics. Opportunity costs calculated from projected farm revenues will, however, be overtaken in the long run by the larger cost of being forever left behind in the technology race. This is a world in which there are promoters, scientists, technology owners, early adopters, follow-on adopters and the fish-eating public. Only the swift-moving early-adopter farmers are likely to gain any longterm competitive advantage from new technology.

Large-scale commercial or national genetics programs tend to benefit the public, professional geneticists and the owners of the technology. Economic benefits to individual farmers who are competing with each other in a commodity market are less obvious unless they become early adopters and/or own part of the technology.

The situation which confronts farmers is summed up in Alice's conversation with the Red Queen in the Land of the Looking Glass: " 'Well, in our country', said Alice, still panting a little, 'you'd generally get to somewhere else - if you ran very fast for a long time as we've been doing.' 'A slow sort of country!' said the Queen. 'Now, here, you see, it takes all the running you can do, to keep in the same place. If you want to get somewhere else, you must run at least twice as fast as that!"

Increased Growth rate Through Classical Broodstock Selection

Mass selection has been by far the most commonly used artificial selection procedure in aquaculture (Hulata, 2001). Hulata's comprehensive review gives instances of successful mass selection performed on many aquacultural species. Even in common species like Atlantic salmon and tilapia, however, the procedure is sometimes successful (Sanchez et al., 1995) and sometimes not (Teichert-Coddington and Smitherman, 1988; Hulata et al., 1986). The frequent failure of well-conducted mass selection programs in aquaculture has yet to be satisfactorily explained.

It is certain, however, that hatcheries that adopt a 'breed the biggest' approach to mass selection for growth soon encounter practical problems arising from variation in spawning times and condition of the breeders. The biggest animals may be somewhat older, rather than faster growing, and may then have this initial, non-genetic, size advantage magnified by competition to such an extent that selection for size becomes ineffective. Straight-forward mass selection often does work, however. In the shrimp Penaeus stylirostris, Goyard et al. (2002) showed that a straight-forward mass selection program can produce useful genetic gains in growth rate. Selection intensities, which ranged between 4% and 18%, were not especially high. The program produced a 21% increase in growth rate by the fifth generation nevertheless.

Hetzel et al. (1999) developed two breeding lines in another species of shrimp, P. japonicus, one mass-selected for high weight and one for low weight. The direct response to this selection was an 8.3% gain in weight of the offspring of the high line. The low line lost about 13%. The authors conclude that the realised heritability of growth rate in mass selection is only moderate but the large family sizes and phenotypic variability (opportunity for selection) should permit rapid stock improvement. These remarks apply to all aquacultural fish and shellfish.

Surprisingly, the least successful selection programs tend to have been those which went well beyond simple mass selection in the sophistication of their statistical, quantitative genetic methodologies (i.e. index selection using the so-called 'animal model').

Several of the most widely promoted projects, continuing for decades in salmon and tilapia, have yet to publish any peer-reviewed proof of genetic gain. Successful index selection has in fact been described in the peer-reviewed literature only twice, to my knowledge (Hershberger et al., 1989; O'Flynn et al., 1999). The realised response to selection in those experiments was not, however, what the authors had predicted from their prior heritability estimates. Selection projects that have worked best in aquaculture have up to now been based on the simpler experimental designs.

It is not at all clear what has gone wrong with sophisticated index selection along classical lines. One possibility is that, as we now know, the genetic architecture of quantitative traits differs from the assumptions made when selection indices are calculated. The genetic basis of the variation in quantitative traits probably involves a few genes having large effects and many genes having small effect (Orr, 1999), contrary to the assumptions of the 'infinitesimal model' from which classical heritability estimates are usually derived.

Competition Effects in Sophisticated Selection Designs

A more likely explanation, specific to aquaculture, are the statistical difficulties caused by competition among individual fish within a tank, cage or pond. It has been known for a very long time that competition can have drastic effects both on the estimation of genetic parameters and the outcome of selection programs (Doyle and Talbot, 1986; Jobling, 1983; Moav and Wohlfarth, 1974; Purdom, 1974). Estimates of genetic parameters such as heritabilities and genetic correlations, which are used in sophisticated breeding programs, are distorted because individual error terms in the statistical analyses are not independent, as they are assumed to be in the linear statistical models used to estimate them (Hamblin and Rosielle, 1978; Mazer and Schick, 1991).

The food that one animal eats is not available to others; small individuals may be intimidated by others even when food is abundant, etc. The genetic variance component due to the genotypes of other individuals in the population is hidden from ordinary genetic analysis. Wolf (2003) found a negative covariance between direct and indirect (competitive) genetic effects, such that genes that make an individual bigger make other individuals smaller. This effect is surprisingly large, and furthermore it increases as the genetic relatedness of individuals in the competing group increases.

Thirty years ago Moav and Wohlfarth (1974) pointed out that in common carp non-genetic differences in size due to an age difference of only one day become magnified by competition and can dominate the outcome of a selection program. At the present time, however, the lamentable truth is that roughly 99.9% of aquaculture genetic analyses ignore competition despite the fact that it has major nongenetic effects on the variance of growth rate. There are, of course, exceptions, including the work of Moav and his colleagues in Israel. Another recent exception is the work of Brichette et al. (2001) on competitive growth of mussels grown in trays. The competition problem may disappear in index selection as selection incorporates information from more relatives extending over more generations.

Since the competition effect on index and combined selection has not been well studied mathematically, we do not know whether to expect that happy outcome or not. Information derived from DNA markers may perhaps be used to generate the required pedigree structure extending over several generations in populations that are grown and propagated together, as is often the case in aquaculture, but as yet no multigeneration selection experiments based on pedigree markers have been published. The accuracy of pedigree reconstruction from markers is turning out not to be especially high (Thomas et al., 2002; Coltman and Slate, 2003; Wilson et al., 2003).

The between-family component of a selection index is always suspect because of non-genetic, environmental effects common to members of a family even when families are reared together in a common tank and DNA markers used to sort them out (e.g. Cunningham et al., 2001; Fishback et al., 2002). The idea that animals in the same tank experience the same environment is an illusion; a family added to the tank on the last spawning day will have a smaller average size than a family added on the first day and will therefore be at a competitive disadvantage. It will also be relatively underfed if feed is supplied to the tank at a rate which is determined by the size and weight-specific metabolic rate averaged across all families.

Within-Family Selection

Successful artificial selection programs in aquaculture tend to be those in which the effects of un-analysed competition are minimised operationally (e.g., by selection within families (Uraiwan and Doyle, 1986) rather than statistically. In pure within-family selection, the biggest animals in each (usually fullsib) family are selected as breeders (Hill et al., 1996).

Mean differences among families, such as age effects, which can be magnified by competition, are ignored. Because aquacultural species usually produce large broods, selection intensities within families can be high. Bolivar and Newkirk

(2002) described a withinfamily selection experiment in tilapia, which achieved a realised heritability estimate of only 0.14 over 12 generations. However, within-family selection permitted high selection intensities of about 2%. The result was that the size of the fish at 16 weeks more than doubled after 12 generations.

Families must be distinguishable in within-family selection, either by growing them in separate tanks until they are big enough to be tagged or by using DNA markers for identifying families. The latter procedure, called 'walk-back' (Doyle and Herbinger, 1994), maximizes effective population size and can achieve very high selection intensities in aquaculture because of the large fecundities. The potential benefit of this procedure over ordinary within-family selection is that families do not have to be reared separately until they are big enough to be physically tagged.

Selection among families (as opposed to within families) is probably only useful when the selected trait cannot be directly measured on the individual chosen to become a breeder. In practice, this probably limits its utility to selection for resistance to or tolerance of disease when challenge tests are used to identify superior families. Non-challenged and therefore non-infected siblings are used as breeders (e.g. Argue et al., 2002; Henryon et al., 2002; Oliver et al., 2000; Sarder et al., 2001).

Chapter 5

Biological Limitations on Transgenic Growth

Transgenics stands far above all other productivityenhancing genetic technology in terms of potential payoff and risk. However, there are some preliminary indications that the most direct transgenic procedure, inserting a transgenic growth hormone, may be relatively ineffective in lines which have already been selected for high growth (Devlin et al., 2001; Parks et al., 2000). It appears that selective improvement and transgenesis may not combine additively. To the extent that this turns out to be true in general, selected strains that achieve growth rates comparable to transgenics may remain competitive with transgenics, with much lower development costs. The effect of a transgene on growth may be strongly influenced by the genetic background of the host.

Devlin et al. (2001) compared the effect of a growth hormone transgene in slow-growing, wild rainbow trout with its effect in rainbow trout that had a long prior history of domestication and selection for fast growth. They comment that "the growth response is strongly influenced by the intrinsic growth rate and genetic background of the host strain, and that inserting growth-hormone transgenes into highly domesticated fish does not necessarily result in further growth enhancement." The growth of the transgenic fish speeded up 17-fold (!) but was still not faster than that of the highly domesticated strain. The domesticated strain hardly responded to the transgene at all. Cranial abnormalities were seen in the transgenic but not in the domesticated animals, which were growing at about the same rate, suggesting that ordinary homeostatic mechanisms were not coping with the novel pathways of growth and development induced by the transgene.

Devlin and his co-authors conclude that "The effect of introducing a growth-hormone gene construct into fish to increase growth rates appears to be dependent on the degree to which earlier enhancement has been achieved by traditional genetic selection. Such effects are likely to be specific for different species, strains and transgenes — in selected mice or in domesticated, rapidly growing farm animals, for example, growthhormone transgenesis can have little effect on growth or it can induce pathological effects, as we have seen in transgenic salmonids."

Aquaculture geneticists, like other biologists, may be able to learn a lot from the study of mice. Bunger and Hill (1999) selected lines of mice for high and low body weight for more than 50 generations, after which the high and low lines had diverged approximately 3-fold in their weight at 98 days. The authors then eliminated growth hormone from the metabolism of the mice by genetic 'knock out', which they achieved by backcrossing a defective growth hormone (GH) releasing factor receptor gene into both lines. Control high and low lines with the normal GH gene were also maintained.

Both lines of mice carrying the knock-out gene, which were thereby deficient in GH, were much smaller than the normal control mice at 98 days. There is no doubt that growth hormone makes mice grow quickly. What is surprising is that the divergence of the high and low lines was almost as great in the absence of growth hormone (2.4-fold divergence) as in its presence (3.1-fold). The authors conclude that after appropriate scale transformation, "changes in the GH system contribute only a small part of the selection response in growth ... [and] other systems contributed most of the selection response". This experiment should interest the aquaculture community even though it was performed on mice. We know that transgenic fish carrying extra growth hormone genes, or modified genes that express GH continuously, are fast-growing fish - sometimes very fast-growing. This ingenious knock-out experiment on mice is a hint that the converse may not be true.

Selection of fast-growing fish by classical methods may evoke an entirely different kind of genetic change that does not involve growth hormone. Furthermore, it suggests that if crosses between highand low-selected lines are used in searches for growth QTLs, the growth hormone system will not necessarily provide the best candidate genes.

Transgenic Disease Resistance

Resistance to disease is of particular interest in the culture of salmonids and shrimp, a fact which is reflected in the focus of ongoing projects in genetic engineering. Standard techniques for inserting foreign genes have been difficult to apply to shrimp because embryos of Penaeus are released from their mothers at a relatively advanced stage. Newlyfertilized eggs are essentially unavailable at the appropriate stage for microinjection or electroporation. Sarmasik et al. (2001) may have found a way around this problem, which is expected to work in other crustaceans and live-bearing fish.

The foreign gene is carried into the host by an extensively engineered viral vector. One engineered feature of the vector makes it unable to replicate. Other features, derived from the hepatitis B virus and the vesicular stomatitis virus (a pathogen similar to hoof and mouth disease which infects mammals, insects and possibly plants), enable the vector to stick to the cell membrane of a wide variety of organisms. Immature gonads of the crayfish were injected with a solution of the vector about one month before the normal age of first reproduction. When they matured the injected individuals were mated with normal individuals. Sarmasik et al. (2001) provide proof of integration, expression and transmission of the reporter transgene for at least three generations.

Some of the more promising transgenes for disease resistance are genes encoding lectin molecules. Lectins are small peptides (amino acid sequences) that bind to sugar molecules exposed on the surface of cell membranes. After binding, some types of lectin lyse the phospholipid bilayer of the membrane, killing the cells. Lytic peptides are proving to be potent toxins to a broad range of bacterial, fungal and protozoan pathogens. Much work has gone into producing transgenic plants and mice that express enhanced levels of lectins as built-in fungicides, bactericides and insecticides. Disease-related transgenic experiments in aquaculture have focused on Cecropin-B, an antimicrobial peptide of about 35 amino acids which is synthesized in the pupae of the silk moth in response to bacterial infection. Electroporation has been used to incorporate cecropin-producing genes into Medaka, with a resulting increase in resistance to Pseudomonas fluorescens, Aeromonas hydrophila, and Vibrio anguillarum (Sarmasik et al., 2002). Challenge studies showed that while about 40% of the controls were killed by both pathogens, only up to 10% of the F2 transgenic Medaka were killed by P. fluorescens and about 10% to 30% by V. anguillarum.

A similar transgene construct and insertion procedure greatly increased the resistance of the channel catfish Ictalurus punctatus to the epizootic of Flavobacterium columnare in an earthen pond (Dunham et al., 2002). Fully 100% of the transgenic catfish survived a natural exposure to the flavobacterium, versus 27% survival of normal fish. When challenged in tanks with Edwardsiella ictaluri, a bacterium that causes enteric septicaemia in catfish, survival of the transgenic fish was 41% versus 15% for the controls.

The use of cecropin transgene in aquaculture has been patented (Cooper and Enright, 1999). The patent claims that "Augmentation of the host's defences against infectious diseases or tumours is achieved by "arming" the host's cells with an exogenous gene encoding a natural or synthetic lytic peptide. ...The transformed cells have the ability to produce and secrete a broad spectrum chemotherapeutic agent that has a systemic effect on certain pathogens, particularly pathogens that might otherwise evade or overcome host defences."

Marker-Assisted and Qtl Selection

The search for disease-resistance and other loci that have effects which are large enough to be useful but not large enough to be obvious by simple segregation analysis (quantitative trait loci, or QTLs) is following two approaches, marker-assisted selection (MAS) and the search for 'candidate genes'. Both approaches are expected to be most useful when ordinary quantitative genetic procedures for estimating breeding values have especially low accuracy – in particular, when selecting for disease resistance if the selected animals cannot be exposed to the disease. MAS and QTL selection will be less useful in selecting for growth where heritabilities are usually reasonably high and estimation of the breeding values of individuals from their phenotypes is reasonably accurate. Simulation studies of selection on growth rate, a trait that is measured on all animals prior to selection, find only small gains from addition of marker data (Lande and Thompson, 1990).

A typical example of MAS for a disease-related trait in aquaculture is provided by the work of Ozaki et al. (2001) on QTLs associated with susceptibility to infectious pancreatic necrosis virus (IPNV) in rainbow trout (Oncorhynchus mykiss). Backcrosses between resistant and susceptible strains were used to identify several chromosome regions containing putative QTL genes that affect disease resistance. Fifty-one microsatellite markers were used for the linkage analysis.

Perry et al. (2001) employed a rather similar approach in finding a QTL for upper thermal tolerance in outbred strains of rainbow trout. Segregation at the microsatellite marker for the QTL explained 7.5% of the variance in thermal tolerance in the trout progenies. This is about what we would expect for a relatively large QTL.

The candidate gene approach to finding QTLs involves looking for genetic variants in biochemical or developmental pathways that are known, or strongly suspected, to affect the trait of interest. An example is the work of Schulte et al. (2000) on the lactate dehydrogenase-B gene (Ldh-B) in northern and southern populations of the fish Fundulus heteroclitus. Northern (Newfoundland) fish grow better at lower temperatures while fish from Florida are superior at higher temperatures. The experimental details are too complicated to be easily summarized here but they include temporary transgenesis of the regulatory sequences into the livers of experimental fish; deletion studies to identify the approximate location within the regulatory sequence where the adaptive changes in the transcript occurred; stress tests of live fish to see which alleles (northern or southern) drive the transcription of the gene. A difference of only one base pair in the regulatory sequence accounts for the adaptive difference between the northern and southern populations.

It appears that over the long term, phenotypic selection for quantitative traits may give better results than either MAS or QTL selection. In a simulation study Villanueva et al. (2002) found that selecting for one particular gene allows the other additive 'background' genes to drop out of the population by chance (inbreeding rate will usually be an indicator of this effect). Thus there is a loss of additive genetic variance with QTL and MAS selection relative to phenotypic selection, and the ultimate selection limits are lower. Interestingly, the reverse seemed not to happen in the simulation - only very rarely was the advantageous QTL allele lost during phenotypic selection. However, in the short term, both QTL and MAS gave a more rapid initial response than phenotypic selection. Salmon and some other aquacultural species have such long generation intervals that rapid response could actually be worth more, by an economic calculation like net present value, than a high selection plateau that might not be approached for 100 years.

The Major Histocompatibility Complex and Disease Resistance

Interest in genetic variation in the major histocompatibility complex of fish (MHC) is running high these days. The diversity of the MHC loci, which are the foundation of vertebrate immune systems, appears to be driven by the diversity of pathogens in the environment (Penn et al., 2002). Fish may choose their mates to optimize the MHC genotype of their offspring (Landry et al., 2001). The preferred explanation for this 'disassortative mating', in which fish choose mates which are genetically unlike themselves, is that MHC heterozygotes are intrinsically more fit than homozygotes (overdominant selection at MHC loci; Arkush et al., 2002). The thought is that heterozygosity at MHC loci may enhance a host animal's resistance to pathogens by increasing both the diversity of peptide antigens it presents to T-cells and the diversity of the T-cells themselves.

But there is also evidence (Miller et al., 2001) that particular MHC alleles may be directionally selected, i.e. towards homozygosity, possibly on a lake-specific basis. Heterozygotes would not be more fit than MHC homozygotes and disassortative mating should not be selected in such lakes. And in an experiment in an aquaculture-like environment where exposure to specific pathogen strains was controlled, particular MHC alleles appeared to have a selective advantage but heterozygosity did not (Lohm et al., 2002).

Along with laboratory, field and theoretical studies of the advantages of immune-system diversity per se, there have also been demonstrations that specific pathogens can exert strong selection on particular MHC alleles in fish. Lohm et al. (2002) reported on the resistance to furunculosis in Atlantic salmon originating from the Akvaforsk strain currently reared by AquaGen AS in Norway. Families of salmon were mated on the basis of their MHC genotypes so as to generate mixtures of homozygous and heterozygous individuals within the same full-sib families, thus controlling the genetic background against which specific MHC alleles were expressed. Surprisingly, in this controlled breeding experiment MHC heterozygosity per se did not improve resistance to the furunculosis challenge test. It was particular Class II MHC alleles that conferred relative fitness differences as great as 0.5. The authors write, "This study clearly shows a strong survival advantage for individuals carrying a high-resistance allele when exposed to a bacterial infection. ... directional selection acting on the MHC despite its high polymorphism stresses the importance of renewal of genetic variation at these kinds of loci, either from mutation, recombination or immigration from other populations, when combating new or coevolving virulent pathogens."

A recent study by Cohen (2002), which delved deeper into the molecular structure of the MHC antigen-binding sites, could be a harbinger of more powerful ways to investigate and exploit MHC variation in fish. A population of Fundulus heteroclitus was found to have adapted to an environment which has been grossly polluted with PCBs and other contaminants for more than half a century and which is toxic to other Fundulus. The population also tolerates high loads of parasites (helminths and others) which are rare or absent in other populations. By studying this and control populations living in more benign environments the authors found amino acid substitutions which tend to be concentrated in different parts of the antigenbinding region of the molecule. They proved that the MHC variation is driven by selection, not drift. The first step in applying this technique in other situations, e.g. searching for selectable QTLs in an aquaculture broodstock, would seem to be finding a population that is unusually well adapted to the targeted stress.

We would like to know what the best genetic management strategy is: selection for homozygosity of particular MHC alleles, or selection for MHC diversity per se. The answer is important both to conservationists and to geneticists who hope to profit from the development of proprietary 'super breeds' for aquaculture. Evidence from other organisms (e.g. mice; Penn et al., 2002) suggests that it probably depends on the variety and timing of challenges anticipated from pathogens. Optimal selection strategies for populations growing in extensive and 'biosecure' aquaculture systems may be even more different than we thought. MHC diversity considerations may become crucial, both practically and politically, in the design of captive breeding and aquaculture genetics programs (Arkush et al., 2002).

Diversion of Resources Away from Growth and into Reproduction

Early-maturing male salmon and trout are a problem for aquaculture because male fish are physically unappealing and their growth slow. Age-at maturation is genetically the same trait in both sexes in salmon, judging by the strong genetic correlation between the males and females for both age of maturation and weight (Kause et al., 2003). Thus it will not be easy to develop a strain in which males mature late but maturation of females is unchanged. The heritability of both traits is sufficiently high, though, that selection for late maturation of both sexes should work if performed on either sex.

There is an analogous problem in tilapia with the difference that the female, not the male, diverts resources towards reproduction at the expense of growth. An ingenious genetic work-around for sexual dimorphism in tilapia has been found and is achieving considerable commercial success. As described by Mair et al. (1997), the procedure involves five preparatory generations of progeny testing and hormonal sex reversal, both male-to-female and the reverse. The final result is a set of YY 'supermales' which, when mated with normal XX females, produce offspring which are nearly 100% normal XY males. The YY supermales have a few female offspring, however, presumably because of the multifactor sex determination in this species. YY male breeders are now used commercially in many places to generate grow-out populations that consist entirely of genotypically normal XY males. The lack of females contributes to uniformity and more rapid growth and also stops unwanted reproduction in aquaculture ponds.

The only major problem with the above procedure is that it interferes with selection for traits such as growth rate (in the YY male donor line, at least) because of the five generations of preparation. The lag could be reduced considerably if the genotypic sexes could be identified without progeny testing. There is reason to hope that commercially useful sex-specific markers can be found in tilapia. Harvey et al. (2002) report the development of in situ hybridization probes to identify sex-specific sequence differences in the long arm of chromosome 1 of the tilapia species Oreochromis niloticus. The binding difference between the probe sequences from X and Y chromosomes is small, which is not surprising.

The authors comment that "Only limited sequence divergence between the X and Y chromosomes would be expected as YY individuals can develop into males or, if hormone treated, females that are both viable and fertile, although growth and survival rates are somewhat lower in YY than XY males.... This suggests that only a very limited loss of function can have occurred in Y-linked genes and that sequence differences between the X and Y chromosomes are largely confined to non-coding regions."

Lee et al. (2003) used bulked segregant (BS) analysis to search for microsatellite marker genes associated with phenotypic sex in tilapia. (In BS analysis DNA from many individuals with the same phenotype is pooled and compared to a pool of DNA from a contrasting phenotype. The contrast here was male vs female phenotypes.) Ten markers were found, all on linkage group 8, which is therefore the (or a) putative Y-chromosome. The linkage of two markers with the sex-determining region was so tight that the sex of offspring of two families was correctly predicted 95% of the time. Unfortunately the markers were not linked to sex in the third family, which shows that we are still some distance away from using markers instead of progeny testing in the commercial production of tilapia YY-supermales.

The Existential Dilemma of Aquaculture Farmers

Farmers beset with problems of cost, price and risk are faced with a dilemma because ignoring genetics is likely to be just as costly as paying for it. Practical decisions about genetics must be made every time fish or shellfish are stocked. Are these breeders the best available? Are they are even average? Would one of the breeds promoted on the worldwide web grow well enough to stave off bankruptcy? The application of modern genetics is often listed as one of the highest priorities in aquaculture research and development agendas. Nevertheless, farmer associations and government agencies are slow to invest in 'big genetics', as has often been noted and lamented by professional geneticists. This is not because 'big business' is attempting to direct publicsector research towards its own interests, as it is reputed to have done during the development of scientific agriculture (Kloppenburg, 1988). Individual farmers just seem to prefer activities that they can do themselves on their own farms.

The question a farmer asks about programs that promise to provide 'super fish' or 'super shrimp' is which program will bring the largest economic benefit to his own farm. As indicated earlier, modern genetics has shown itself capable of producing genetic productivity gains measured not by increments but by multiples of the existing standard. Every time a farmer stocks or re-stocks his farm he realises the existential horror of his predicament: the available genetic information is irrelevant and incomplete in its most crucial practical aspects; doing nothing about genetics is also taking action; the appropriate emotions for the practical person are anguish and dread.

When a new agricultural technology is introduced everyone involved can be considered either a beneficiary, a bystander or a victim. The immediate beneficiaries are easy to identify. Professional geneticists (some of them) will benefit from being the heroes of the technology and so will their research sponsors. The owners of the technology when it is successfully commercialized will also benefit. The fish-eating public will, eventually, be able to buy fish at a lower price.

Some farmers - the early adopter farmers - will also obtain a competitive commercial advantage which benefits them in the short term. The same competition, however, victimizes follow-on farmers in the short term. Simple economic theory predicts that farm-gate prices will go down, production will be up and farmers will become increasingly dependent on technology over which they have no control, just as in modern terrestrial agriculture (Allen, 1984; Kloppenburg, 1988; Weller, 1999). The owners of the technology and the fish-eating public will be sharing the benefits of the higher productivity permitted by the new technology and farmers will be in the same cost-price squeeze as they were before - essentially, bystanders.

If a new strain of fish or shellfish is available to all producers, market competition occurs and a new equilibrium price is reached at the new and lower production cost which, however, now includes whatever extra fees the producers have to pay for the new strain.

It will of course benefit the technology developers to make their technology universally available – for a price - and not restrict it to a few producers. The benefits of the increased production efficiency accrue to the public, which may pay less for fish, and to the sellers of the technology. Competition for market share again requires producers to reduce their price and profits to the lowest tolerable level.

The above reasoning leads to the conclusion that farmers who want to benefit from purveyors of advanced technology should try to become early adopters and possibly insist on some sort of exclusive franchise arrangement to protect their competitive advantage as long as possible. The farmer should also attempt to obtain sole or part ownership of the technology in order to continue benefiting as bioeconomic equilibrium is approached. The developers of the technology will be anxious to attract early adopters for promotional reasons and the leverage this gives the farmer might be translated into franchises or ownership. It should be emphasized that there is a divergence of interest between the developers of the technology and the farmers who use it. The former group wishes to maximise the spread of the new technology, for reasons of professional prestige (including institutional prestige in the case of universities, governments and international development agencies) as well as commercial gain.

The Dilemma of Aquaculture Genetics Companies

James (2000) recently distinguished between three types of genomics companies: product providers, information providers and technology providers. Although he wrote about companies that are working on human health issues, his analysis applies equally well to genetics in aquaculture. Companies in these three areas are now racing for primacy and moving into each other's commercial strategy space. The question which interests Mr. James is how to invest money. The question which interests aquaculturists is how to bet the future of the fish farm, since a bet on genetics must be made.

If we apply James's analysis to aquaculture genetics we conclude that companies which provide proprietary products like vaccines or genetically improved broodstock and fingerlings can potentially make the highest profit but also experience the highest risk, in particular the risk that someone else will develop a product which is cheaper or more effective. Purveyors of aquaculture genetic information about genomic sequences, markers and maps are mostly but not entirely in the public sector. The opportunity to generate value from proprietary information about QTLs, pathogens and broodstock genotype environment interaction in aquaculture is not being ignored, however. The commercial risk to information suppliers is that their proprietary information will become 'commoditised' and freely available. Some people are even making a moral crusade out of the public right to raw genetic data. Spider Robinson neatly summed it up in the Toronto Globe & Mail on 18 March, 2000: "It's as though an explorer took the first photo of a zebra - then claimed ownership of zebras, the concept of stripedness, and anything else substantially zebraic in nature".

Both in human genomics and in aquaculture genetics there are companies that develop technology for use by other companies e.g. for on-farm broodstock improvement. (My own consulting company falls into this category.) James notes that such companies "though in some ways offering the lowest risk for investors, are always in danger of becoming generic or outdated as new ways for tackling a problem are developed."

Given the entirely rational preference that farmers have for technologies over which they have sole ownership, professional geneticists could usefully focus more effort on developing high- as well as lowtechnology procedures that can be applied on a small scale. Farmer-breeders would then have more scientific support and information when choosing how to act at the farm level. They might then feel more comfortable in their existential dilemma where any procedure for choosing breeders, even random choice, has genetic consequences. Such procedures would enable individuals to develop proprietary breeds on their own farms and should speed up the application of modern genetics to aquaculture.

Genetic Selection Programme of Fishes

There are two ways a fish farmer can increase fish production: The first is to increase the size of the fish farm. The second is to increase yield, which is the weight of fish produced in each pond. It is often impossible to increase the size of the farm, because either the land is unavailable, the land is prohibitively expensive, or the land that can be used or water supply associated with that land is so poor that the costs of production would exceed the value of the crop. Even when a farmer can increase his land holdings, it often makes sense to improve production efficiency and produce greater yields (kg/ha). There are two ways a fish farmer can increase yields: The first is by environmental manipulations, such as the increased use of lime, fertilizers, feeds, and/or improved water quality management. The second is by growing genetically improved fish. If both approaches are used, yields can increase dramatically.

There are a number of breeding programmes that can be used to improve a population of fish genetically. Selective breeding and crossbreeding (also called "hybridization") are the two traditional approaches that have been used for thousands of years, and they have been used to improve all major crops and livestock grown by farmers. Inbreeding is often combined with hybridization to improve the results of the crossbreeding programme. Three additional approaches have been developed recently: chromosomal manipulation, production of sex-reversed brood stock, and genetic engineering.

Although farmers have conducted breeding programmes on livestock for thousands of years, fish farmers are only beginning to use selection, hybridization, or other breeding programmes to improve aquacultured species of food fish. Although some progress has already been made, many fish farmers are culturing fish that are essentially wild and unimproved.

Selective breeding is a breeding programme that tries to improve the breeding value of the population by selecting and mating only the best fish (largest, heaviest, those with the desired colour, etc.) in the hope that the select brood fish will be able to transmit their superiority to their offspring. If this occurs, the next generation will be more valuable because the fish will grow faster, which will increase yields; the fish will grow more efficiently, which will lower feed costs; or all fish will have a more desired body colour, which will increase their market value.

Crossbreeding is a breeding programme that tries to find mating combinations between different populations of fish which produce superior offspring for grow-out, offspring that are said to exhibit hybrid vigour. Although crossbreeding is a tried-and-true method of increasing yields, the results of crossbreeding programmes are impossible to predict (unless the mating has been made previously), so the production of superior offspring is a hit-or-miss proposition. Many combinations often have to be evaluated before one is discovered that produces offspring with hybrid vigour. Crossbreeding programmes usually involve different strains within a species (intraspecific hybridization), but different species can also be hybridized (interspecific hybridization). To date, much of the breeding work in aquaculture has been devoted to hybridization among the different species of tilapia in an attempt to produce all-male hybrids for grow-out. In general, crossbreeding is used to produce superior fish for grow-out (production fish), while selection is used to create superior brood fish. The hybrids that are created in a crossbreeding programme are usually grown and sold as food. A farmer rarely retains and spawns hybrids to produce a new generation of production fish. On the other hand, select brood fish that are created in a selective breeding programme are created for one purpose-to produce the next generation of fish for grow-out-and their offspring can, in turn, be retained and selected to continue the process.

In recent years, biotechnological research has led to the development of three additional breeding programmes that can be used to increase yields. One of the most common breeding programmes in aquaculture is the production of sex-reversed brood stock to produce monosex populations for grow-out. This is done either because one sex is superior or more desirable or to prevent reproduction during grow-out. For example: female sturgeon are more valuable than males because they produce caviar; female salmon are the more valuable sex, because sexually precocious males die before they can be harvested; male tilapia are more desired than females because they grow twice as fast. The major goal in tilapia farming is to prevent reproduction during grow-out; this can be best accomplished by producing a monosex male population.

The production of sex-reversed brood stock is usually accomplished by feeding either estrogens or androgens (sex hormones) to sexually undifferentiated fry to sex-reverse them. Sex-reversed fish are individuals that are one sex phenotypically but the other genetically. If sex reversal is done properly, sex-reversed fish are capable of producing monosex populations for grow-out. The type of hormone used-estrogens to produce sex-reversed females or androgens to produce sex-reversed males-depends on the sex-determining system of the species and whether you want to produce an all male or an all-female population.

Another breeding programme that is becoming more commonplace is chromosomal manipulation. This is usually done to produce sterile fish. The most common form of chromosomal manipulation is to use temperature or pressure to shock newly fertilized eggs (chemicals can be used, but they are less effective). If this is done properly, the shock prevents the second polar body from leaving the egg, so the newly fertilized egg contains a haploid sperm nucleus, a haploid egg nucleus, and a haploid second polar body nucleus. These three haploid nuclei fuse and produce a triploid zygote which, in turn, produces a triploid fish. Triploids are sterile. This type of breeding programme is used to enable farmers to grow exotic species whose culture might otherwise be illegal or to induce sterility in species that become sexually mature before they reach market size. For example, grass carp culture in most of the United States is legal only if a farmer raises triploids. This technique can also be used to improve the results of interspecific hybridization.

Chromosomal manipulation can be used to produce animals that have only a mother (gynogens) or only a father (androgens). This is done by creating haploid zygotes and by then shocking the zygotes to produce diploid zygotes. Haploid zygotes are produced in one of two ways: a normal egg is fertilized by sperm whose DNA has been destroyed by UV irradiation (gynogenesis); a normal sperm is used to fertilize an egg whose DNA has been destroyed by UV irradiation (androgenesis). Gynogenesis and androgenesis are techniques that can be used to produce highly inbred lines for breeding purposes. It can also be used to produce supermales; such males are capable of producing all-male populations.

In recent years, a new, high-tech breeding programme has been developed: genetic engineering. This is a breeding programme that transfers a single gene or a set of genes from one individual into another. This transfer can be within a species, between two species, or even across kingdoms. Although genetic engineering has generated lots of publicity, to date it has not produced genetically superior fish for farmers. Furthermore, this type of breeding programme is very expensive, highly regulated, and requires highly trained scientists. This type of breeding programme should be conducted only by scientists working at universities, at governmental research stations, or at agribusinesses that are capable of supporting expensive research projects with secure containment facilities. Although all breeding approaches that can be used to improve yields are important and although they can be used either singly or in combination to achieve specific goals, this manual will describe only those procedures that can be used to improve fish by selective breeding.

The decision to conduct a selective breeding programme is a decision that must be made for each farmer or each fry/fingerling production centre on a case-by-case basis. The decision to incorporate selective breeding into a farmer's work plan should not be a general decision made at a regional level. If it is, most of the selective breeding programmes will be failures, because selective breeding programmes require dedication, a certain level of sophistication, record keeping, and the investment of extra labour. Additionally, selective breeding programmes are not free; they also require the investment of money. Finally, these programmes usually do not produce immediate improvements. Improvements are usually not seen for at least one growing season, so a farmer must be able to incorporate long-term planning into his farm management programme, and he must be patient. As a result, within a region, only a small percentage of farmers or fingerling production centres should or will ever conduct selective breeding programmes.

A final requirement that must be met before a farmer can conduct a selective breeding programme is the existence of proper facilities. This manual outlines selective breeding programmes that can be used to improve growth rate and other phenotypes on a medium-sized farm or fingerling production centre, which was arbitrarily defined as a farm with about 2 ha of ponds. The breeding programmes outlined in this manual can be conducted in 1–150 ponds, depending on the type selection that will be used and the culture system that is typically used to grow the fish. Additional ponds will be needed to hold brood fish and to spawn them; other types of facilities, such as holding tanks, will also be needed.

Finally, common sense must prevail when choosing the most appropriate breeding programme. Even if a farmer has the ability and the desire to conduct a selective breeding programme, the biology of the species and the way it is grown should be carefully considered before the decision is made to conduct a selective breeding programme to improve growth rate. Even though most farmers would like to have faster-growing fish, in some cases greater yields can be achieved by improving other phenotypes via another type of breeding programme. For example, the biggest problem in tilapia culture is the fact that tilapia become sexually mature before they reach market size and, as a result, reproduce in the grow-out ponds. This uncontrolled reproduction means that a significant percentage of yield is unmarketable. Tilapia farmers may benefit from breeding programmes that can produce monosex male populations far more than from selective breeding programmes that might improve growth rate.

Genetics

The ultimate goal of every selective breeding programme is to improve the breeding value of the population. The breeding value is determined by the fish's genes. A farmer hopes that when he improves the breeding value of his population he will also improve its monetary value, which is determined by the fish's phenotypes.

To accomplish this goal, a breeder selects (saves) fish that possess certain phenotypes and culls (removes) those that do not, and hopes that his breeding programme will create a genetically improved population (breeding is the applied science of genetics; thus, a farmer who conducts a selective breeding programme is a breeder). If it does, the fish in the next generation will be more valuable because their genes will enable them to grow faster or to exhibit a more desirable colour.

Although the goal of a selective breeding programme is to manipulate a population's genes and thus produce better fish, it is impossible to examine and manipulate the genes directly. Instead, a fish's genes are examined indirectly by examining its phenotypes (also called "traits"), which are the physical expressions of the genes. Because the goal of a selective breeding programme is to manipulate a fish's genes, it is important to understand how the genes are transmitted from a parent to its offspring and how the genes produce the phenotypes. Understanding these processes helps explain how selection works. This is important, because if a farmer understands why he is doing something, he is more likely to do it correctly, and this ensures success. Additionally, an understanding of how the genes produce various phenotypes enables a breeder to choose the breeding programme that will allow him to achieve his goal quickly and efficiently.

This chapter is not an all-encompassing review of basic genetics. Only a few topics are covered: The first section discusses meiosis, which is the process by which a parent's genes are parcelled into its gametes (eggs and sperm).

The second section describes the difference between phenotype and genotype. The remaining sections explain how phenotypes are inherited.

The information contained in this chapter is important because many of the terms introduced in this chapter will be used in Chapters 3 and 4 to explain how selective breeding programmes should be conducted and how they work. It is not necessary to become a geneticist in order to conduct a successful selective breeding programme, but a basic understanding of how phenotypes are inherited will make the rest of the manual easier to understand and will also enable a farmer to be a better breeder.

The material in this chapter is intended simply as background information for those who seek a better understanding of how selection alters phenotypes and improves a population. The material in this chapter, especially that dealing with quantitative phenotypes, is intended for extension agents and for highly educated aquaculturists. It is not intended for the vast majority of farmers or for those who are interested only in the nuts and bolts of how to conduct a selective breeding programme and who do not care how it works.

It is hoped that extension agents will read this material, because it will provide a better grounding in fish breeding, and an understanding of how selection works genetically may make them more comfortable with the design and implementation of a selective breeding programme.

Those who are not interested in this material or those who already understand basic genetics can skip this chapter and go directly to Chapters 3 or 4 which describe selective breeding programmes.

Meiosis

Genes are located on structures called "chromosomes," which are located in the nucleus of every cell. Although there are some exceptions, for the most part, chromosomes occur in pairs, which means that a gene is a paired entity. Throughout this manual, we will assume that chromosomes in cultured fishes occur in pairs. One chromosome of each pair comes from a fish's mother, while the other comes from its father. The number of chromosomes varies among species, but it is constant within a species.

Although every cell in a fish contains the entire genome (a fish's genetic make-up—every gene on every chromosome), the genes that exist in the primary gametocytes are those that are ultimately of greatest interest to a breeder. All genes in the various cells of a fish are, of course, important because they produce the various phenotypes that breeders work with and ultimately select. But those in the primary gametocytes are of greater importance because the primary gametocytes are the cells which develop into sperm and eggs-the gametes which carry the genes and which produce the next generation of fish.

Primary gametocytes are the cells that develop into eggs and sperm by a process called "meiosis." Meiosis is one of the most important of all biological processes, because it greatly increases genetic variability through crossing over and through independent assortment. In addition, mistakes that occur during the replication of chromosomes in meiosis are heritable, and this is how new alleles and genes are created. The end result of meiosis is the creation of haploid sperm and eggs. Were it not for meiosis, life would not have evolved past single-celled organisms.

During the initial part of meiosis, the chromosomes replicate, and the homologues of each pair come together (the homologs are the two chromosomes that form a pair; one homolog comes from the father, and one comes from the mother). The replication of the homologues is usually uneventful, in that they replicate themselves perfectly. Occasionally, however, a mistake is made during replication, and a gene is not replicated perfectly. Such mistakes are called "mutations." The mutation rate for individual genes is rare and ranges from 1 in 10,000 replications to 1 in 100,000 replications. Although rare, the fact that males are capable of producing several hundred million to over a billion sperm and that females can produce tens of thousands to hundreds of thousands of eggs all but guarantees that each fish will produce dozens of gametes which contain one or more mutations. The production of mutations is a very important biological process, because it increases genetic variability by creating new alleles and new genes which can, in turn, produce new phenotypes.

Because each homolog replicated before they paired, the bundles are composed of four chromosomes and are called "tetrads." The chromosomes in each tetrad are extremely elongated and drawn out rather than compact, and they twist around each other. As they twist, the chromosomes break, and pieces from different homologues rejoin. The exchange of genes from one homolog to another is called "crossing over," and it is one of the most important genetic events, because this greatly increases genetic variability by creating new and different gene combinations every generation. These different gene combinations will, in turn, increase phenotypic variability, which helps breeders produce better fish.

The second important process that occurs during meiosis is reduction division. During this process, the chromosome number is reduced from the diploid (paired state) state to the haploid (unpaired) state. When chromosomes are paired, a cell or individual is said to be a diploid (2N). When a cell or individual contains only one chromosome from each pair, it is said to be a haploid (N).

During reduction division, the replicated homologues of each chromosome pair separate, and the primary gametocyte then divides and forms two cells. In males, these cells are the secondary spermatocytes; in females, they are the secondary oocyte and the first polar body. The separation of the replicated homologues of each chromosome pair and the direction in which the replicated homologues go is independent of that which occurs for all other chromosomes. Although each tetrad divides itself along parental lines (the replicated homolog that came from a fish's father separates from the replicated homolog that came from its mother), the division of the homologues of each chromosome pair is random and is independent of that which occurs in all other tetrads. The random division of the maternal and paternal chromosomes is called "independent assortment," and it is extremely important because it creates new chromosomal and new gene combinations, which greatly increase genetic variability; this ultimately increases phenotypic variability. The secondary spermatocytes and the secondary oocyte are haploid, since each contains only one homolog from each pair.

The final step in meiosis is the equational division, during which the replicated homologues of each chromosome separate and go into one of two sperm cells or into the egg or second polar body. As was the case during reduction division, the direction that a replicated homolog from each chromosome goes is random and independent of that which occurs in all other chromosomes. This final shuffling of the chromosomes and of the genes also increases genetic variability.

The end result of meiosis is the production of haploid sperm and eggs. Each gamete contains a single chromosome from each pair. Even though some fish can produce millions of gametes, few if any are identical because of mutations, crossing over, and the independent assortment of chromosomes that occurs during reduction division.

The reduction of the chromosome number from the diploid state in the primary oocytes and primary spermatocytes to the haploid state in the eggs and sperm is critical. If this did not occur, the number of chromosomes would double each generation. Because gametes are haploid, the normal diploid number for a species is restored when a sperm fertilizes an egg.

Phenotype and Genotype

A gene or set of genes contains the blueprints or chemical instructions for the production of a protein. This protein either forms or helps produce various phenotypes, such as body colour, sex, number of rays in the dorsal fin, length of a fin, body length, and weight. When a geneticist talks about this process, he says that a fish's genotype controls or produces its phenotype. The genotype is the genetic make-up of the fish. It is the gene or genes that controls a particular phenotype. Because chromosomes occur in pairs, genes also occur as pairs (there are some exceptions, but they will not be discussed in this manual), so the genotype is a paired entity.

A gene can occur in more than one form. Alternate forms of a gene are called "alleles." In a population, a gene may exist in only one form, which means that there is only one allele at a given locus (locus = gene), or there may be up to a dozen alleles at a locus.

Because chromosomes occur in pairs, an individual can possess either one or two alleles at a given locus. Even if there are ten alleles for a specific gene in a population, an individual can possess no more than two (in this manual, we will assume that all fish are diploids). If the pair of alleles at a given locus is identical, the fish is said to be "homozygous" at that locus. If the pair is not identical, the fish is said to be "heterozygous" at that locus. The terms homozygous and heterozygous (the genotype) refer to specific genes, not the fish's entire genome. A fish's genome is composed of tens of thousands of genes, which is a mixture of homozygous and heterozygous loci.

The reason why it is important to make a distinction between individuals that are homozygous or heterozygous, is that different forms of a gene (alleles) produce different forms of that gene's protein. This means that the various alleles at a locus are responsible for the production of various body colours or of different rates of growth. And it is these differences that interest geneticists, because they can be exploited by selection to produce faster-growing or more attractive fish.

The phenotype is the physical expression of what the gene or set of genes produce, and this is what we describe (for example, colour or sex) or measure (for example, length or weight). Breeders divide phenotypes into two major categories: qualitative phenotypes and quantitative phenotypes.

Qualitative Phenotypes Produced by Single Autosomal Genes

Most qualitative phenotypes that have been deciphered genetically in food fish are controlled by single autosomal genes with two alleles per locus. In general, genes express themselves either in an additive or in a non-additive manner. In additive gene action, each allele contributes equally to the production of the phenotypes in a stepwise unidirectional manner, and the heterozygous phenotype is intermediate between the two homozygous phenotypes. In non-additive gene action, one allele (the dominant allele) is expressed more strongly than the other (the recessive allele), and it has a greater influence on the production of the phenotypes.

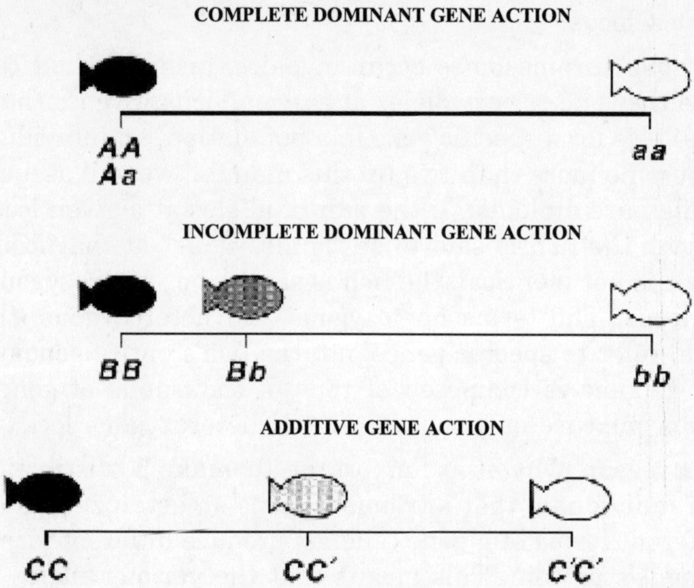

Figure: *Schematic diagram of qualitative phenotypes that are controlled by a single autosomal gene with either complete dominant gene action, incomplete dominant gene action, or additive gene action. Genotypes are given below the phenotypes. Gene A produces black and white colours by complete dominance, so there only two phenotypes. Black is the dominant phenotype, and it is produced by both the homozygous dominant (AA) and heterozygous (Aa) genotypes. White is the recessive phenotype, and it is produced by the homozygous recessive (aa) genotype. Gene B produces black and white colours by incomplete dominance. Because the mode of gene action is incomplete dominance, the heterozygous (Bb) genotype produces a unique phenotype (light-black), one that resembles but that is slightly different than the dominant phenotype (black), which is produced by the homozygous dominant (BB) genotype; white is the recessive phenotype, and it is produced by the recessive genotype (bb). Gene C produces black and white colours by additive gene action. Because neither allele is*

dominant, the heterozygous (CC') genotype produces a unique phenotype (gray) that is intermediate between the phenotypes (black and white) produced by the two homozygous genotypes (CC produces black and C'C' produces white). When the mode of gene action is additive, there is no dominant or recessive allele or phenotype.

Complete Dominant Gene Action: Complete dominant gene action occurs when the dominant allele is so strong that it produces its phenotype, regardless of the genotype. Only a single dominant allele is needed to produce the dominant phenotype.

This means the homozygous dominant and heterozygous genotypes both produce the dominant phenotype; thus, the phenotypes produced by these genotypes are identical.

The recessive allele can produce the recessive phenotype only when no dominant allele is present, which means it can produce the recessive phenotype only when a fish is homozygous recessive.

Consequently, with this mode of inheritance, there are three genotypes but only two phenotypes:

Phenotype	***Genotype***
Dominant	Homozygous dominant
Dominant	Heterozygous
Recessive	Homozygous recessive

For example, light-coloured (pink) and normal pigmentation phenotypes in Nile tilapia are controlled by the B gene. The dominant B allele produces the dominant normally pigmented phenotype, while the recessive b allele produces the recessive pink phenotype. Because the B gene exhibits complete dominance, normal pigmentation is produced by both the homozygous dominant genotype (BB) and by the heterozygous genotype (Bb). while pink is produced only by the homozygous recessive genotype (bb). Table 1 lists some qualitative phenotypes in important cultured food fishes that are produced by single autosomal genes with complete dominance.

Table: *Examples of phenotypes in cultured food fishes that are controlled by single autosomal genes with complete dominant gene action. All phenotypes in this table are body colours except caudal deformity syndrome, which is a tail deformity.*

Species	Recessive phenotype	Dominant phenotype
Common carp	blue	normal pigmentation
	gold	normal pigmentation
	grey	normal pigmentation
	normal pigmentation	light yellow band on dorsal fin; yellow on head
Nile tilapia	blond	normal pigmentation
	syrup	normal pigmentation
	light-coloured (pink)	normal pigmentation
	caudal deformity syndrome	normal tail
	normal pigmentation	red
Grass carp	albino	normal pigmentation
Channel catfish	albino	normal pigmentation
Rainbow trout	albino	normal pigmentation
	iridescent metallic blue	normal pigmentation

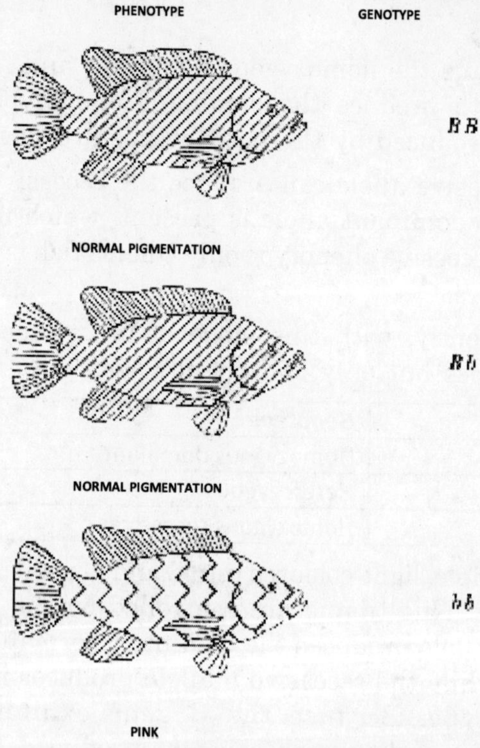

Figure: *Inheritance of normally pigmented and pink body colours in Nile tilapia. These phenotypes are controlled by a single autosomal gene with complete dominant gene action called the B gene: the dominant B allele produces normal pigmentation, while the recessive b allele produces pink. Because the B allele is completely dominant over the b allele, the BB and Bb genotypes both produce the dominant normally pigmented phenotype. The recessive pink phenotype is produced only when a fish is homozygous recessive (bb). The pictorial representations of body colour used in this figure will also be used in Figures 3 and 9.*

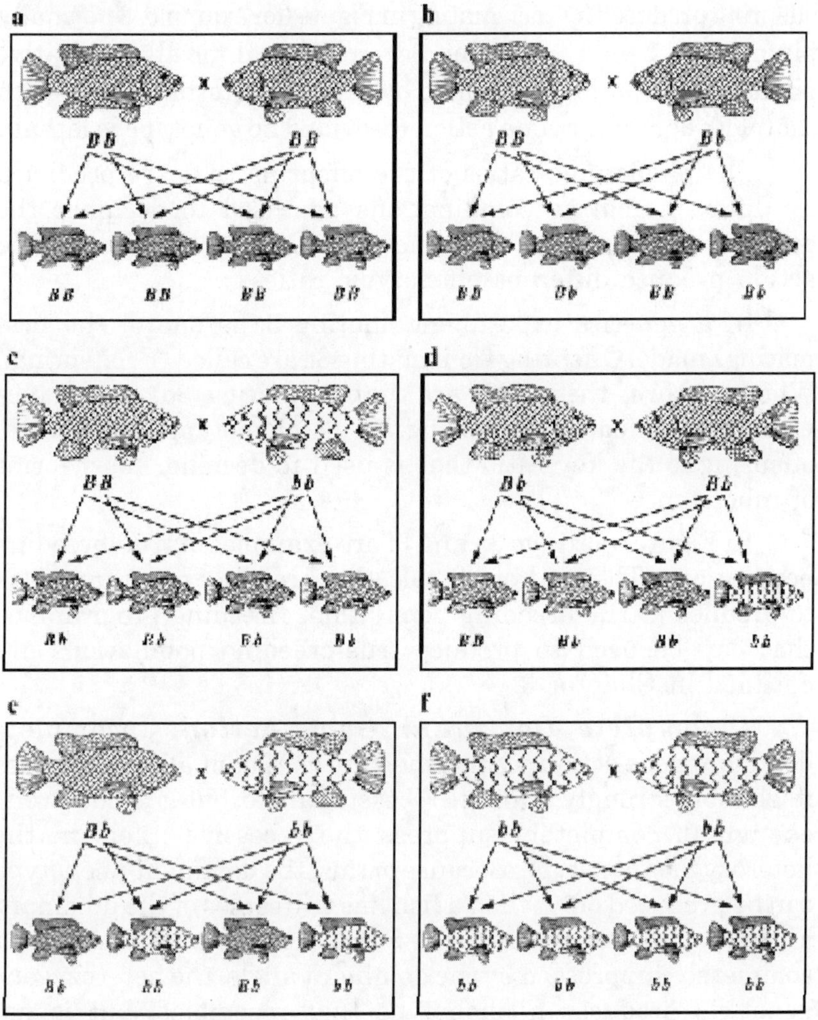

Figure: *All possible mating combinations between normally pigmented and pink Nile tilapia, and the phenotypes of the offspring produced by each mating. Pictorial representations of the phenotypes are the same as those used in Figure. Genotypes are given below the fish. The arrows represent gametes. Mating combinations a and f are examples of true-breeding populations, the goal of a selective breeding programme.*

Figure depicted earlier illustrates all possible mating combinations between pink and normally pigmented Nile tilapia and the phenotypes of the offspring produced by each mating. The mating combinations and the phenotypic ratio of the offspring produced by each mating in Figure (for example, 3 normally pigmented: 1 pink in mating 3d) are typical for all qualitative phenotypes that are controlled by a single autosomal gene with complete dominant gene action (provided no genotype is lethal).

The phenotypic ratios of the offspring that are produced by different mating combinations are used to decipher the genetics that controls the phenotypes. Different modes of gene action produce different phenotypic ratios.

In a genetics experiment, mating 3c is one of the first matings made. Offspring from mating 3c are called "F_1 offspring." When mature, these fish are mated (mating 3d) to produce what are called "F_2 offspring." The phenotypic ratio of F_2 offspring is the key ratio that is used to decipher most forms of inheritance.

In Figure, matings 3a and 3f are examples of true-breeding populations. The goal of all selective breeding programmes is to produce a true-breeding population. Breeding programmes that can be used to produce true-breeding populations are outlined in Chapter 3.

***Incomplete Dominant Gene Action*:** Incomplete dominant gene action occurs when the dominant allele expresses itself morc strongly than the recessive allele, but it is not strong enough to completely suppress the recessive allele in the heterozygous genotype. Because of this, the dominant phenotype can be produced only when a fish has two copies of the dominant allele (homozygous dominant). Since the recessive allele is not completely suppressed by the dominant allele, the heterozygous genotype produces a phenotype that resembles, but is not identical to, the dominant phenotype. As was the case with complete dominance, the recessive phenotype is produced only when a fish is homozygous recessive. Because the heterozygous genotype produces a phenotype that is similar to but is different from the dominant phenotype, when the mode of gene action is incomplete dominance, there are three genotypes and three phenotypes, a unique phenotype for each genotype:

Phenotype	*Genotype*
Dominant	Homozygous dominant
Dominant	Heterozygous
Recessive	Homozygous recessive

For example, black (normal pigmentation), bronze, and gold body colours in Mozambique tilapia are controlled by the G gene. The dominant G allele produces melanistic (dark-coloured) fish, but because the G gene exhibits incomplete dominance, the G allele does not completely suppress the expression of the recessive g. allele in the heterozygous state.

The homozygous dominant and heterozygous genotypes produce unique phenotypes: GG fish are black, while Gg. fish are bronze. Gold fish are produced by the homozygous recessive genotype (gg.). Table 2 provides examples of qualitative phenotypes in important cultured food fishes that are produced by single autosomal genes with incomplete dominance.

Figure illustrates all possible mating combinations among black, bronze, and gold Mozambique tilapia and the phenotypes of the offspring produced by each mating.

The mating combinations and the phenotypic ratio of the offspring produced by each mating in Figure (for example, 1 black:2 bronze: 1 gold in mating 5d) are typical for all qualitative phenotypes that are controlled by a single autosomal gene with incomplete dominant gene action (provided no genotype is lethal). In Figure, matings 5a and 5f are examples of true-breeding populations, and the phenotypic ratio of the offspring produced by mating 5d is the key that is used to unlock the mode of inheritance.

Additive Gene Action: With only one exception, all qualitative phenotypes controlled by a single autosomal gene that have been discovered in cultured food fish are controlled either by complete or by incomplete dominance. The lone exception is a gene that controls golden, palomino (the heterozygous phenotype) and normal body colours by additive gene action in rainbow trout.

When phenotypes are controlled by additive gene action, there is no dominant or recessive allele. Both alleles contribute equally to the production of the phenotypes, so the heterozygous genotype produces a phenotype that is intermediate between those produced by the two homozygous genotypes. Consequently, when the mode of inheritance is additive gene action, there are three genotypes and three phenotypes, a unique phenotype for each genotype. The difference between additive gene action and incomplete dominance is that when the mode of gene action is incomplete dominance, the heterozygous phenotype approximates one of the homozygous phenotypes (the dominant one), while it is intermediate between the two homozygous phenotypes when the mode of gene action is additive.

Because qualitative phenotypes are descriptive and are not measured, it is possible to mis-classify incomplete dominance and additive gene action.

Even if the mode of gene action is mis-classified, it is of no practical importance for a breeding programme. Because both types of gene action have three genotypes and each produces a unique phenotype, the breeding programmes that are used to control phenotypes produced either by incomplete dominance or by additive gene action are identical.

The mating combinations and phenotypic ratio of the offspring produced by the matings that are shown in Figure can also be used to illustrate what occurs for phenotypes that are controlled by a single autosomal gene with additive gene action.

The phenotypic ratios for phenotypes controlled by incomplete dominance and by additive gene action are identical. The only difference is the appearance of the heterozygous phenotype-does it resemble the dominant phenotype (incomplete dominance), or is it intermediate between the two homozygous phenotypes (additive)?

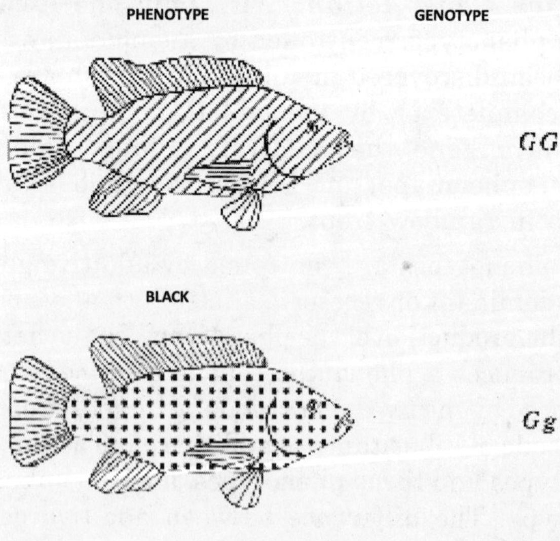

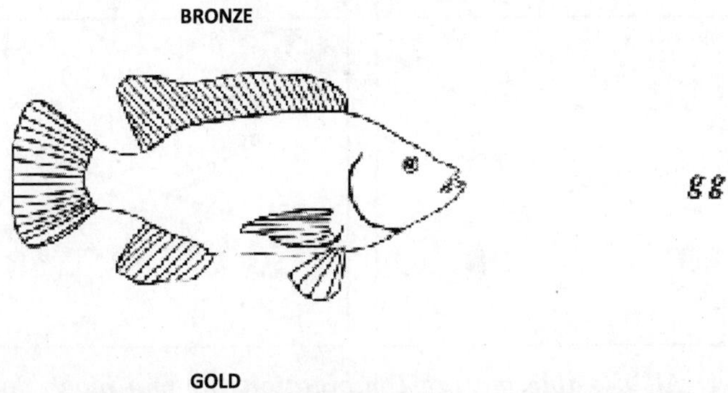

Figure: *Inheritance of black, bronze, and gold body colours in Mozambique tilapia. These phenotypes are controlled by a single autosomal gene with incomplete dominant gene action called the G gene. Because the dominant G allele is not completely dominant over the recessive g. allele, the heterozygous genotype produces a phenotype that is similar to but distinct from that produced by the homozygous dominant genotype. Homozygous dominant fish (GG) are black (the dominant phenotype); heterozygous fish (Gg) are bronze (the heterozygous phenotype); homozygous recessive fish (gg) are gold (the recessive phenotype). Pictorial representations of body colour used in this figure will also be used in Figures 5, 10, 11, 17, and 18.*

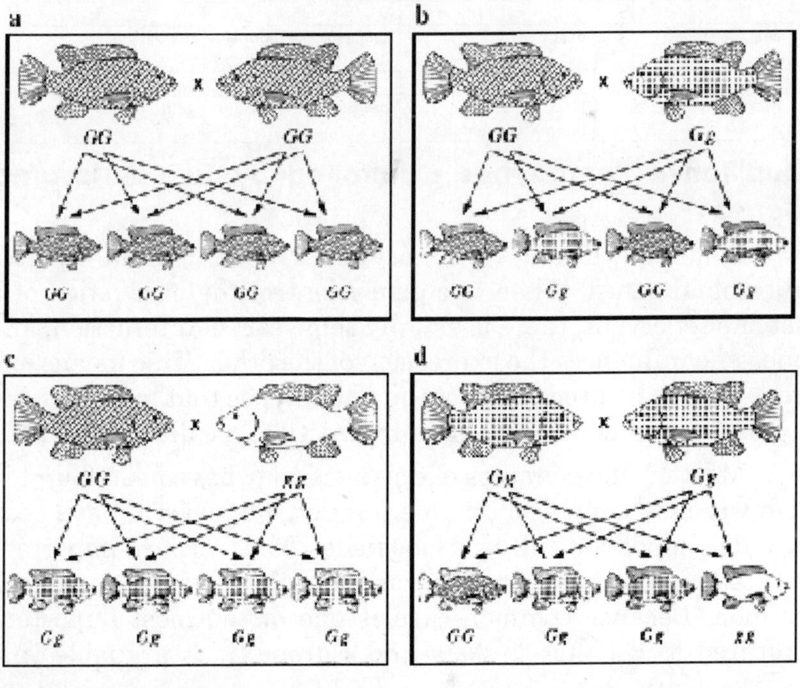

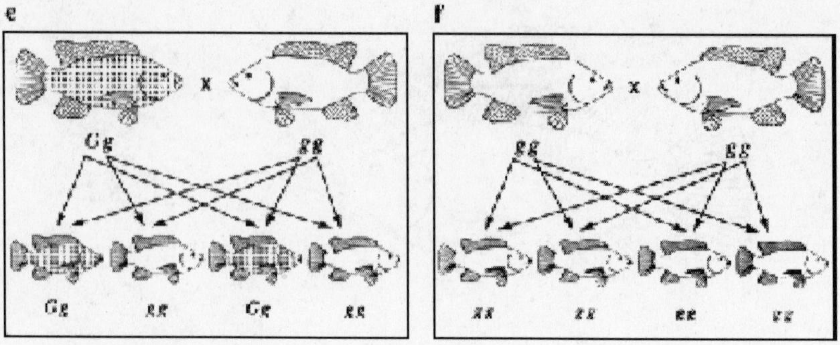

Figure: *All possible mating combinations among black, bronze, and gold Mozambique tilapia, and the phenotypes of the offspring produced by each mating. Pictorial representations of the phenotypes are the same as those used in Figure. Genotypes are given below the fish. The arrows represent gametes. Mating combinations a and f are examples of true-breeding populations, the goal of a selective breeding programme.*

Table: *Examples of phenotypes in cultured food fishes that are controlled by single autosomal genes with incomplete dominant gene action.*

Species	***Dominant phenotype***	***Heterozygous phenotype***	***Recessive phenotype***
Common carp	death	light coloured	normal pigmentation
Blue tilapia	death	saddleback (abnormal dorsal fin)	normal
Mozambique tilapia	black (normal pigmentation)	bronze	gold

Qualitative Phenotypes Controlled by Two Autosomal Genes

Some qualitative phenotypes are controlled by two autosomal genes. When two genes control the production of a set of phenotypes, there is usually some sort of interaction, and one gene influences the expression of the other. This means one gene alters the production of the phenotypes that are produced by the second gene. This gene interaction is called "epistasis."

Most of the examples of epistasis that have been found in fish were discovered in ornamental fish, but several have been found in important cultured food fishes. The two most important are scale pattern in common carp and flesh colour in chinook salmon. Because common carp is one of the most important cultured food fishes in Asia and Europe, it is arguable that scale pattern in common carp is the most important set of qualitative phenotypes that have been found in any aquacultured species. Scale pattern also helps determine colour pattern, and thus the value, of ornamental common carp (koi).

The four scale pattern phenotypes in common carp-scaled (normal scale pattern), mirror, line, and leather-are controlled by two genes (S and N) with what is called "dominant epistasis." The S gene determines the basic scale pattern via complete dominance. The dominant S allele produces the scaled phenotype (SS and Ss genotypes), while the recessive s allele produces the reduced scale phenotype called "mirror" (ss genotype). The N. gene modifies the phenotypes produced by the S gene. There are two alleles at the N locus.

The dominant N allele modifies the phenotypes as follows: in the homozygous state (NN). the N allele kills the embryo; in the heterozygous (Nn) state, the N allele changes the scaled phenotype into the line phenotype and changes the mirror phenotype into the leather phenotype. The recessive n allele has no effect on the phenotypes produced by the S gene. The five phenotypes (one is death) and the underlying genetics are illustrated.

A set of qualitative phenotypes may be controlled by more than two genes. Body colour in the Siamese fighting fish is an example of a set of phenotypes that is controlled by the epistatic interaction among four genes. Working with these phenotypes is far more complicated because of the number of genes involved. Fortunately, in food fish, no qualitative phenotype controlled by more than two genes has been discovered.

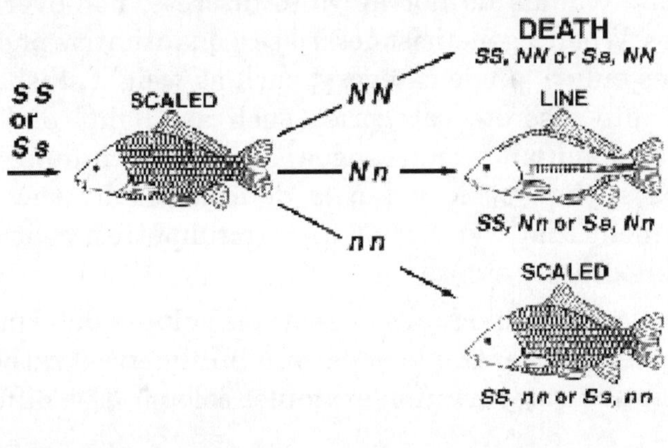

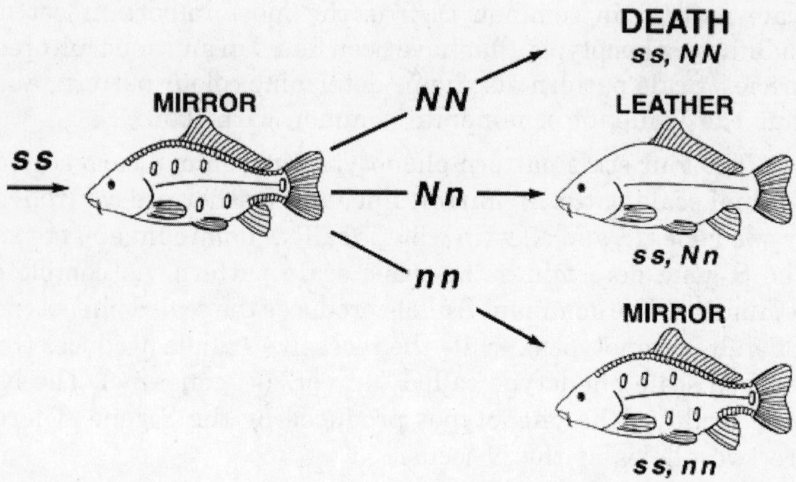

Figure: Inheritance of scale pattern in common carp. Scale pattern is determined by the epistatic interaction between the S and N genes. The S gene determines whether the fish has the scaled phenotype (SS and Ss genotypes) or the mirror phenotype (ss genotype). The N gene modifies those phenotypes. The NN genotype kills the fish (SS, NN, Ss, NN, and ss, NN genotypes); the Nn genotype changes the scaled phenotype into the line phenotype (SS, Nn and Ss, Nn genotypes) and changes the mirror phenotype into the leather phenotype (ss, Nn). The nn genotype does not alter the phenotypes produced by the S gene, so scaled fish have the SS, nn or Ss, nn genotypes, while mirror fish have the ss, nn genotype.

Genetics of Quantitative Phenotypes

Quantitative phenotypes are the phenotypes that are measured, such as length, weight, eggs/kg female, or feed conversion. Quantitative phenotypes differ from qualitative phenotypes in that individuals do not fall into discrete, non-overlapping categories. When a geneticist describes a quantitative phenotype, he creates only a single category, such as weight. Fish are not grouped into discrete categories such as "light" or "heavy." Instead, individuals are arranged along a continuum, and an individual's phenotypic value is determined by the unit of measurement that the farmer uses (millimetres, centimetres, grams, kilograms, etc.).

Because an individual's phenotypic value is determined by measurement (for example, length in millimetres) rather than by descriptive category (for example, colour), the differences between two individuals is a matter of degree (millimetre) rather than of kind (colour). Because the differences among individuals are matters of degree, in a population quantitative phenotypes form what are called continuous distributions, which can be described graphically.

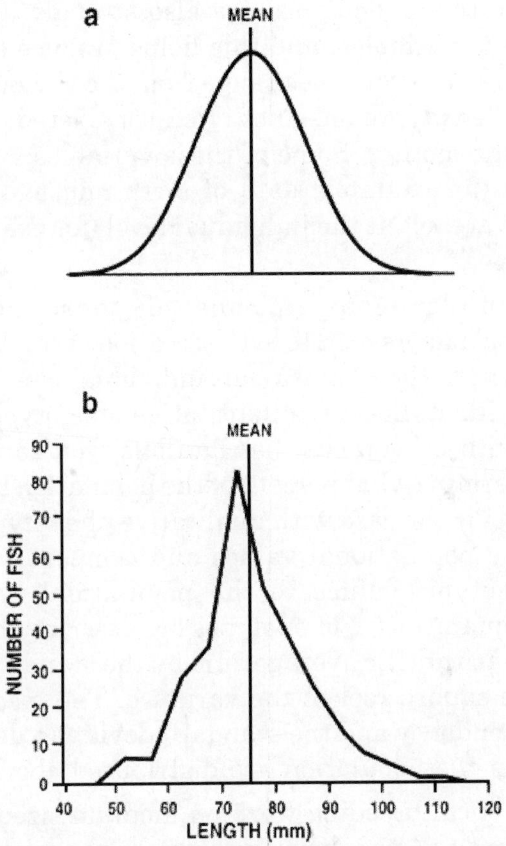

Figure: *Distribution exhibited by a quantitative phenotype in a population. Graph a illustrates a perfect distribution, which creates what is called a "bell-shaped curve" with the mean bisecting the curve at its peak. Graph b is the distribution of 7-month length in a population of common carp.*

The reason why quantitative phenotypes do not segregate individuals into neat and precise categories, as was the case with qualitative phenotypes, is that they are much more complicated genetically. In general, qualitative phenotypes are controlled by one or two genes. They can be controlled by more, but few are. Quantitative phenotypes are controlled by dozens to hundreds of genes. The exact number is usually never known. Additionally, the genes are shuffled like a deck of cards during meiosis due to crossing over and the independent assortment of chromosomes; the combination of these events ensures that each offspring will receive a slightly different genetic message.

Quantitative phenotypes are also strongly influenced by environmental variables, and this helps produce a continuous distribution. These variables range from the obvious ones, such as stocking density, to ones not often considered, such as size and age of the mother. Some of these variables are felt at the family level (for example, date of birth and age of mother), while others are felt at the individual level (for example, access to food).

The simultaneous actions of these genetic and environmental factors results in the creation of single phenotypic categories where the only way an individual can be described is by measuring it. Because quantitative phenotypes are single categories with continuous distributions, you cannot analyse ratios or determine what percent of the population has a specific phenotype, as is the case with qualitative phenotypes. Instead, you calculate populational values and compare individual or family phenotypic values to the population's values. In a population, quantitative phenotypes are described by the mean, which is the arithmetic average, and by the standard deviation, which is the square root of the variance. The mean describes the central tendency and the standard deviation describes how the values in the population are distributed about the mean.

For practical breeding work on medium-sized fish farms, it is important to know how to calculate the mean, so that a farmer can assess the effect of his selective breeding programme. A farmer really does not need to know how to determine the standard deviation; scientists and researchers, on the other hand, must know how to determine the standard deviation. Table 3 illustrates how the mean is determined.

Because quantitative phenotypes are controlled by dozens to hundreds of genes, the simultaneous and/or sequential expression of these genes makes it impossible to identify individual genes and to decipher their modes of inheritance. Consequently, a different approach is needed to work with and to understand these phenotypes. Because quantitative phenotypes are more complicated genetically, it is more difficult to work with these phenotypes, but quantitative phenotypes are the most important phenotypes in agriculture or aquaculture-weight, fecundity, etc.—so the breeding value of a farmed population of food fish is mainly determined by the genes that control quantitative phenotypes.

Their importance is underscored by the fact that quantitative phenotypes are often called "production phenotypes."

***Table**: How to calculate the mean for a quantitative phenotype. In this example, we will calculate mean length. In general, you determine the mean from a random sample of 30–200 fish.*

Step 1.	Obtain individual lengths to the nearest millimeter. Thirty fish are measured									
	98	103	106	111	104	91	87	103	114	107
	101	104	97	105	108	100	110	113	104	105
	95	97	107	108	99	111	112	113	105	103
Step 2.	Determine the sum of the measurements; that is, add the phenotypic values									
	98 + 103 + 106 + 111 + 104 + 91 + 87 + 114 + 103 + 107 + 101 + 104 + 97 + 105 + 108 + 100 + 103 + 113 + 105 + 95 + 97 + 107 + 108 + 99 + 111 + 112 + 105 + 113 + 103 = 3,120									
Step 3.	Divide the total value derived in Step 2 by the number of fish that were measured. In this case, 30 fish were measured, so you divide by 30.									
	$\frac{3,120}{30} = 104$									
	The mean length in this population is 104 mm									

Phenotypic Variance

Because quantitative phenotypes exhibit continuous distributions in a population, the only way to work with and to improve these traits is to analyse their variance and to divide it into the heritable and the non-heritable components. Phenotypic variance is the variability that a phenotype exhibits in a population; the mean describes the average phenotypic value, while the variance describes how individuals are distributed around the mean (the standard deviation, which was mentioned earlier, is the square root of the variance). Phenotypic variance (V_P) is the sum of three components: genetic variance (V_G), environmental variance (V_E), and genetic-environmental interaction variance ($V_{G\text{-}E}$). This can be represented by the following formula:

$$V_P = V_G + V_E + V_{G\text{-}E}$$

Genetic Variance

Obviously, genetic variance is the component that breeders try to manipulate during a breeding programme. Genetic variance is itself the sum of three components, and different breeding programmes are needed to exploit these sub-components.

Genetic variance (V_G) is the sum of additive genetic variance (V_A), dominance genetic variance (V_D), and epistatic genetic variance (V_1). As before, this can be represented by a formula:

$$V_G = V_A + V_D + V_I$$

The words "additive," "dominance," and "epistatic" do not refer to specific types of gene action as they do when discussing the modes of inheritance for qualitative phenotypes. The correct terms are "additive genetic variance," "dominance genetic variance," and "epistatic genetic variance" (not gene action), and they refer to specific components of variance that are produced by the entire genome, not by one or two genes.

Additive genetic variance is the genetic component that is due to the additive effects of all the fish's alleles. One way of thinking about this is that additive genetic variance is the sum of the values that each allele makes to the production of the phenotype. Some alleles will make a large contribution; some will make a small contribution; some will make no contribution; and some may even make a negative contribution. The contribution made by every allele is added, and the sum is the additive genetic variance component for each fish. Dominance genetic variance is the genetic component that is due to the interaction that exists between the pair of alleles at every locus. Because of this, dominance genetic variance cannot be inherited.

The idea that some form of genetics cannot be inherited usually causes confusion, but it is a simple concept. Dominance genetic variance is due to the interaction of the pair of alleles at each locus, which means that dominance genetic variance is a function of the diploid state (2N). The reason that dominance genetic variance is not heritable is that each parent contributes a haploid (N) gamete to the production of each offspring. Gametes do not contain pairs of alleles (2N); the diploid state is reduced to the haploid state during meiosis. During reduction division, all allelic pairs are separated when the chromosomes undergo independent assortment, which means that a parent's dominance genetic effects are destroyed during meiosis. Since an individual's dominance genetic variance is destroyed during reduction division, it cannot be transmitted via a gamete to an offspring-thus, it is not heritable. Dominance genetic variance effects are recreated at fertilization when a haploid sperm fertilizes a haploid egg to produce a diploid zygote. At fertilization, genes once again exist in the paired state, which means that an interaction exists between the pairs of alleles at each locus which, in turn, means that dominance genetic variance exists. Consequently, dominance genetic variance effects are destroyed and then recreated in new and in different combinations each generation.

Epistatic genetic variance is the genetic component that is due to the interaction(s) of alleles between or among loci; in other words, it is the interaction(s) that an allele has with alleles other than its own pair. Epistatic genetic variance is a mixture of heritable and non-heritable variance. The portion of the interaction that is between and among the alleles that were included in a gamete is heritable, but the portion that is between or among alleles that were parcelled to other sperm or to the polar bodies is not heritable. The percentage of epistatic genetic variance that is heritable varies from gamete to gamete because of crossing over and independent assortment. Independent assortment and crossing over tends to disrupt most epistatic genetic variance during meiosis, so only a small random sample is transmitted from a parent to its offspring; consequently, only a small random portion of epistatic genetic variance is heritable.

The differences among additive genetic variance, dominance genetic variance, and epistatic genetic variance and how they are transmitted is important on a practical level because different kinds of breeding programmes are needed to exploit these components of genetic variance. Furthermore, the relative amount of phenotypic variance that can be attributed to these components of genetic variance determines the type of breeding programme that can be used and how effective it will be in improving the phenotype. The two important genetic components are additive genetic variance and dominance genetic variance. Most breeders assume that epistatic genetic variance is not important.

This assumption is made because it is difficult to try and select for combinations of alleles when you do not know what combinations are desirable. Additionally, the improvement that can be gained by selection for epistatic effects is rather small, and it plateaus quickly.

Additive genetic variance and dominance genetic variance are essentially opposites. Additive genetic variance is a function of the alleles, so it is function of the haploid state; dominance genetic variance is a function of allelic pairs, so it is a function of the diploid state. A parent produces haploid gametes, so it can transmit its additive genetic effects to its offspring, but it cannot transmit its dominance effects, which are destroyed during meiosis.

Dominance effects are created in each zygote after fertilization. Thus, the additive effects are a function of each parent, while the dominance effects are a function of specific matings. Because the additive effects are transmitted from a parent to its offspring, additive genetic variance is often called the "variance of breeding values."

Because additive genetic variance is transmitted from a parent to its offspring, selection is the breeding programme that is used to exploit this component of variance and to improve the population.

Because dominance genetic variance is not heritable but a function of the mating, hybridization is the breeding programme that is used to exploit this component of variance and to improve the population.

Heritability: Because additive genetic variance is transmitted from a parent to its offspring in a predictable and reliable manner, if the percentage of phenotypic variance that is due to additive genetic variance is known, a farmer can predict the amount of improvement that can be made as a result of selection, and he can even customize selection to achieve a pre-determined amount of improvement per generation.

The proportionate amount of additive genetic variance is called "heritability," and it can be represented by the following formula:

$$h = V_A/V_P$$

where: h is the symbol for heritability, V_A is additive genetic variance, and V_P is phenotypic variance. Heritability is expressed as a percentage (0–100% or 0.0–1.0). Thus, heritability quantifies the percentage of phenotypic variance that is inherited in a predictable and reliable manner.

The major reason for determining the heritability of a quantitative phenotype is that it can be used to predict the results of a selective breeding programme by using the following formula:

$$R = Sh$$

where: R is the response to selection (gain per generation), S is the selection differential (the superiority of the select brood fish over that of the population average; to determine this, you simply subtract the population average from the average of the select brood fish), and h is heritability. Table 4 shows how a heritability can be used to predict response to selection.

Table: *How to predict the response to selection if the heritability (h) of the phenotype is known. In this example, we will predict response to selection for increased length and then calculate the predicted mean length of the next generation*

Given:	h^2 for length at 12 months = 0.26 mean length at 12 months of the population = 146 mm mean length at 12 months of the select brood fish = 162 mm
Step 1.	Calculate the selection differential (S). S = mean length of the select brood fish - mean length of the population S = 162 mm - 146 mm = 16 mm
Step 2.	Calculate the predicted response to selection (R). $R = Sh^2$ R = (16 mm)(0.26) R = 4.16 mm
Step 3.	Calculate the mean of the F, generation of select fish. mean of the F_1 generation = mean of the population + R mean of the F_1 generation = 146 mm + 4.16 mm = 150.16 mm

The preceding formula clearly demonstrates that heritability is the factor that determines the percentage of selection differential that can be gained via selection; in other words, how much gain is possible. In general, heritabilities e"0.25 indicate that selection will produce good gains, while those d" 0.15 indicate that selection will be ineffective.

Heritabilities > 0.3 are considered to be large. Although it is advantageous to know the heritability of a quantitative phenotype before conducting a selective breeding programme, it is not necessary. If one exists, it can be used to predict the gains, to customize the selection differential that is needed to achieve a desired response to selection, or to indicate that selection will be so ineffective that the programme should be scrapped. Table 5 shows how a heritability can be used to customize the selection differential in order to achieve a desired response to selection. It is often unnecessary to determine a heritability, because published information already exists. Several hundred heritabilities have already been determined for phenotypes such as growth rate, food conversion, disease resistance, fecundity, egg size, egg number, dressing percentage, body conformation, and pesticide tolerance in many important aquacultured species of food fish.

The heritabilities that are published may not be the same as those in a farmer's population, because heritabilities are specific for the population that was evaluated and for the culture conditions that were used in the experiment, but the published values should be similar to those that exist in most populations. Table 6 lists some of the heritabilities that have been determined in common carp and tilapia.

Table: *How to use the heritability (h) of the phenotype to customize the selection differential in order to produce the desired response to selection.*

Given:	h^2 for length at 12 months = 0.26 mean length at 12 months of the population = 146 mm desired response to selection = 6 mm
Step 1.	Calculate the selection differential (S) needed to produce a response of 6 mm:
	$R = Sh^2$ 6 mm = (S)(0.26) 6 mm/0.26 = S 23.08 mm = S
Step 2.	Calculate the mean of the select brood fish that will be needed to produce a selection differential of 23.08 mm.
	S = mean length of the select brood fish - mean length of the population 23.08 mm = mean length of the select brood fish - 146 mm mean length of the select brood fish = 146 mm + 23.08 mm mean length of the select brood fish = 169.08 mm

Selection for Homozygous Phenotypes

When this type of selection is used, it is usually done because a farmer wants to fix the recessive phenotype and eliminate the dominant phenotype, and the phenotypes are produced by an autosomal gene that exhibits complete dominance. The term "recessive" does not imply inferiority; similarly, the term "dominant"does not imply superiority. The terms simply refer to the way the alleles express themselves and to the phenotypes that are produced by those alleles. Many recessive phenotypes have been shown to be quite valuable in various farmed plants and animals, and breeding programmes have been conducted to fix them. The terms "inferior" and "superior" should be applied to phenotypes only after biological and economic evaluations have been conducted.

If the desired trait is the recessive phenotype that is produced by a gene that exhibits complete dominance, the selective breeding programme that is needed to fix the trait and to produce a true-breeding population is simple, and this goal can be achieved by a single act of selection. To accomplish this, fish with the recessive phenotype will be saved, and fish with the dominant phenotype will be culled.

Genetically, fish with the undesired dominant phenotype are either homozygous dominant or heterozygous. Because every fish that has at least one dominant allele expresses the undesired dominant phenotype, a simple one-step culling of all fish with the dominant phenotype will eliminate all copies of the dominant allele. Conversely, every fish that expresses the desired recessive phenotype is homozygous recessive, which means that when these fish are selected, none will possess a copy of the undesired dominant allele. Thus, the recessive allele will be the only allele that exists in the select population, which means that it will breed true. For example, say a tilapia farmer has both normally pigmented and pink Nile tilapia, and decides to grow only pink tilapia. He can create a true-breeding pink population by a simple one-step selective breeding programme during which he will cull all normally pigmented fish. Pink body colour is the recessive phenotype, and it is produced by the recessive b allele (bb genotype); normal body colour is the dominant phenotype, and it is produced by the dominant B allele (BB and Bb genotypes). The culling of all normally pigmented tilapia will remove all B alleles from the population. The only fish that remain will be pink and, since they are homozygous recessive bb, the b allele will be the only allele that will remain in the population of select brood fish. Consequently, a single act of selection will create a true-breeding pink population.

This type of selective breeding programme can also be used to fix either the recessive or the dominant phenotypes when the mode of gene action is incomplete dominance, and it can also be used to fix either homozygous phenotype when the mode of gene action is additive. In both cases, there are two homozygous genotypes that can be identified and isolated. By culling the undesired phenotypes, true-breeding populations can be created by a single act of selection.

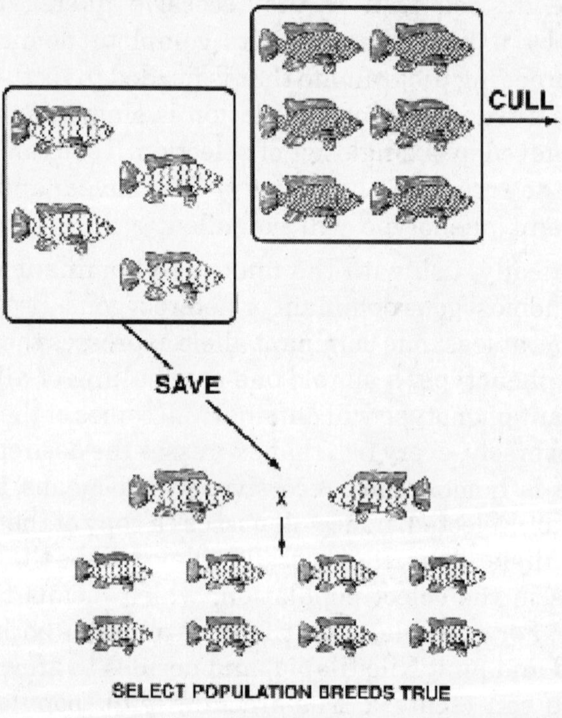

Figure: *Selective breeding programme needed to produce a true-breeding population of pink Nile tilapia. If all normally pigmented (the dominant phenotype) fish are culled, all copies of the dominant B allele will be eliminated. The only fish that remain are pink (the recessive phenotype), and since they are homozygous recessive bb, the select population of pink brood fish will breed true and will produce only pink offspring. Pictorial representations of the phenotypes are the same as those used in Figure. Any recessive phenotype can be fixed this easily*

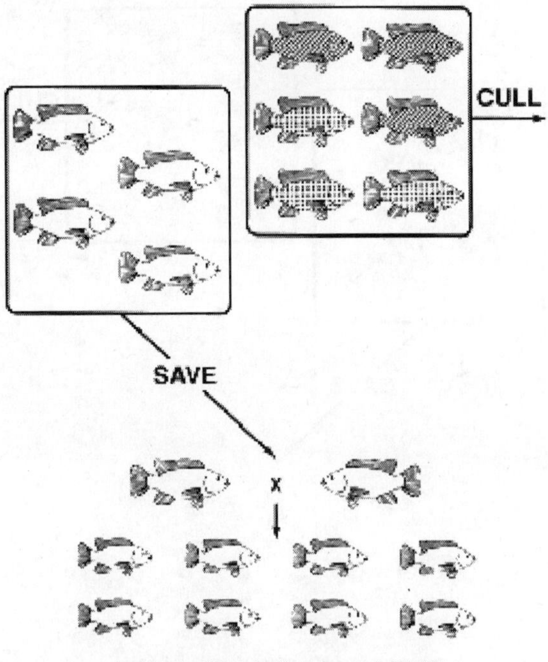

Figure: *Selective breeding programme needed to produce a true-breeding population of gold Mozambique tilapia. If all melanistic (black [GG] and bronze [Ggl]) fish are culled, all copies of the dominant G. allele will be eliminated. The only fish that remain are gold (the recessive phenotype), and since they are homozygous recessive gg, the select population of gold brood fish will breed true and will produce only gold offspring. Pictorial representations of the phenotypes are the same as those used in Figure.*

For example, say a tilapia farmer has black, bronze, and gold body colours in his population of Mozambique tilapia. If the farmer wants to produce either a true-breeding population of gold or a true-breeding population of black fish, he can accomplish either goal by a single act of selection. Gold is the recessive phenotype, and it is produced by the homozygous recessive genotype (gg); black is the dominant phenotype, and it is produced by the homozygous dominant genotype (GG); bronze is the heterozygous (Gg) phenotype.

A true-breeding population of gold tilapia, can be produced by culling all black and all bronze fish. Conversely, a true-breeding population of black tilapia can be produced by culling all gold and all bronze fish.

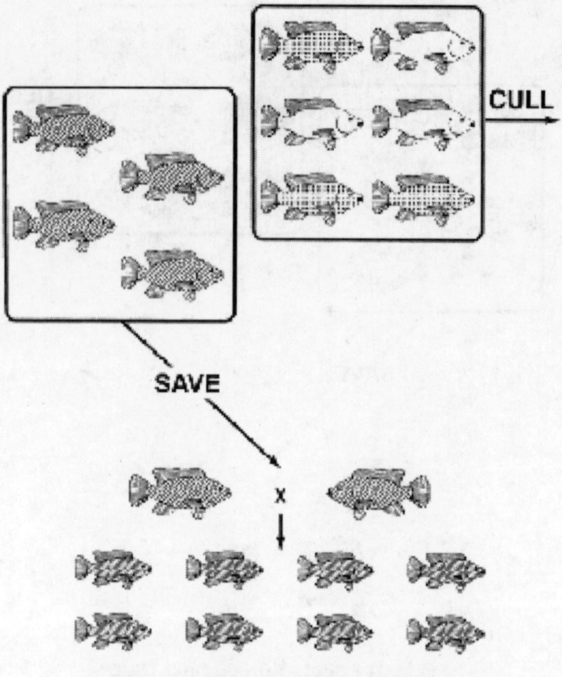

Figure: *Selective breeding programme needed to produce a true-breeding population of black Mozambique tilapia. If all bronze (the heterozygous [Gg] phenotype) and all gold (the recessive [gg] phenotype) fish are culled, all copies of the recessive g. allele will be eliminated. The only fish that remain are black (the dominant phenotype), and since they are homozygous dominant GG, the select population of black brood fish will breed true and will produce only black offspring. Pictorial representations of the phenotypes are the same as those used in Figure.*

This same concept can be extended to phenotypes that are controlled by two or more genes. If the desired phenotype is controlled only by a homozygous genotype, a true-breeding population can be created by a single act of selection.

For example, if a carp farmer has a population of common carp with all four scale phenotypes and wants to produce a true-breeding population that has a reduced scale pattern, all he has to do is save the fish with the mirror phenotype and cull all other fish. Three of the scale phenotypes have a reduced number of scales, but only mirror is produced by a homozygous genotype (ss, nn), so it is the only one that is capable of breeding true. Both the leather (ss, Nn) and line (SS, Nn and Ss, Nn) phenotypes are undesired, because they are heterozygous at the N locus, which means that when two leather and/or line fish mate, 25% of their offspring will die (those that are homozygous NN. regardless of the genotype at the S locus).

Leather and line phenotypes are also undesired, because fish with these phenotypes have several negative pleiotropic effects, among them lowered growth rates and viabilities.

If the farmer culls all leather, line, and scaled individuals, all S and N alleles will be eliminated.

The only fish that will remain in the select population will be those with the mirror phenotype, and since they are homozygous recessive (ss, nn) the s and n alleles will be the only ones that remain in the population of select brood fish.

Consequently, a single act of selection will create a true-breeding population of mirror common carp.

Selection for Phenotypes Controlled by More than One Genotype

Selection cannot fix a phenotype that is controlled by more than one genotype. This means selection cannot fix a dominant phenotype and create a true-breeding population if it is produced by a gene with complete dominance.

Conversely, this also means that selection cannot eliminate a recessive phenotype (and the recessive allele) if it is produced by a gene with complete dominance.

Despite the fact that this type of selective breeding programme will not work, it is routinely used in a misguided effort to eliminate undesired phenotypes that are produced or that are assumed to be produced by recessive alleles.

This type of breeding programme will not achieve its goal for either of two reasons. The first is the fact that most undesired phenotypes are abnormalities that are created by non-genetic factors (environmental disturbances or developmental mistakes).

Many assume that all abnormal phenotypes are mutant phenotypes which are produced by recessive alleles. Some are, but most are not. If the undesired phenotype is produced by a non-genetic factor, selection will not work, because no selective breeding programme can fix or eliminate non-heritable phenotypes.

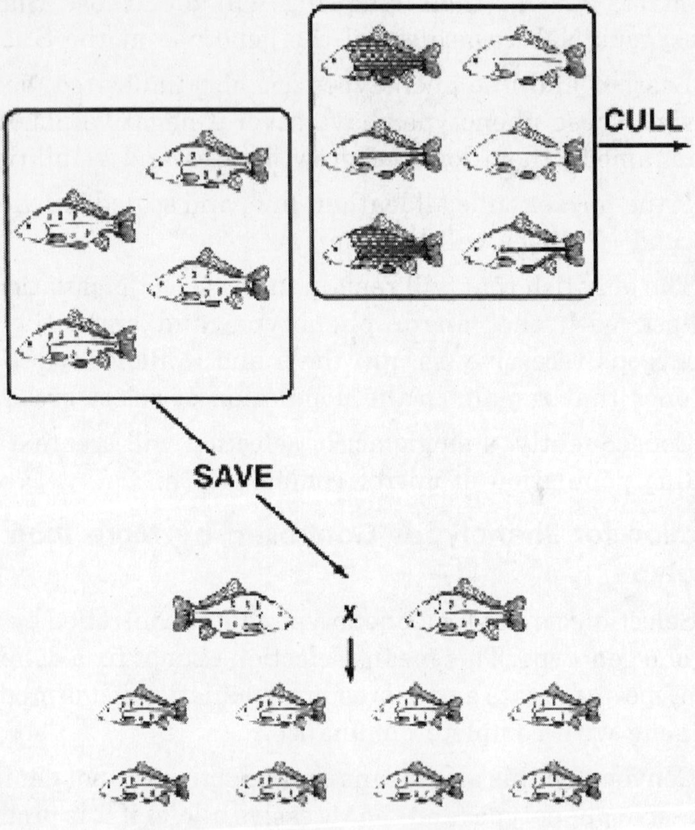

Figure: *Selective breeding programme needed to produce a true-breeding population of mirror common carp. If all scaled, line, and leather fish are culled (phenotypes that are produced by genotypes with at least one dominant allele), all copies of the S and N alleles will be eliminated. The only fish that remain are those with the mirror (the recessive) phenotype, and since they are homozygous recessive ss, nn the select population of mirror brood fish will breed true and will produce offspring with only the mirror phenotype. Even if a set of phenotypes are produced by two or more genes, if the desired phenotype is the recessive phenotype (all loci responsible for the production of the phenotype are homozygous recessive), selection can produce a true-breeding population this easily.*

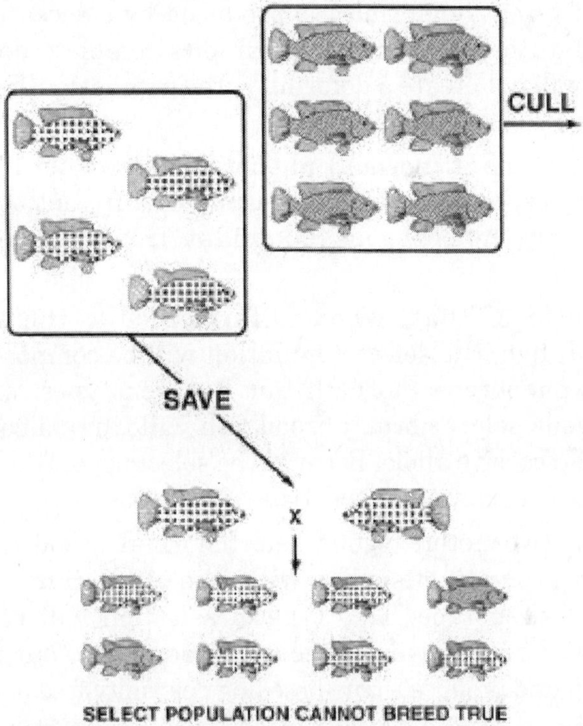

Figure: *Selective breeding programme used in an unsuccessful attempt to produce a true-breeding population of red Nile tilapia. Red fish are represented by fish covered with squares, while normally pigmented fish are those covered with stripes (the same pictorial representations for body colour will be used in Figure 15). If all normally pigmented fish (the recessive phenotype) are culled, the select population will be composed of only red fish (the dominant phenotype). Since red body colour is produced by a dominant allele that exhibits complete dominance, red is produced by both the homozygous dominant (RR) and by the heterozygous (Rr) genotypes. Every heterozygous red brood fish (Rr) that is saved will carry a copy of the r allele to the select population; consequently, selection cannot eliminate it. When select red brood fish are spawned, the mating of two heterozygotes (Rr) will produce some normally pigmented offspring (the mating that is depicted). Recessive phenotypes cannot be eliminated by culling them; they will reappear when the select brood fish are spawned. Consequently, if the desired phenotype is the dominant phenotype and it can be produced by both the homozygous dominant and by the heterozygous genotype, selection cannot create a true-breeding population.*

Even when a deformity is produced by a recessive allele, if it is an autosomal gene that exhibits complete dominance, selection will not create a deformity-free population. The reason why it will fail is quite simple:

The dominant (normal) phenotype is produced by either of two genotypes, and it is impossible to differentiate normal fish that are homozygous dominant from those that are heterozygous.

Because of that, when a farmer culls the recessive (abnormal) fish, the select population will be composed of fish with one phenotype (normal) but two genotypes, and every heterozygous select normal brood fish will carry a copy of the undesired recessive allele. Because the select population contains both alleles, it cannot breed true.

When two heterozygous select normal brood fish mate, they will produce offspring with the undesired abnormal (recessive) phenotype. This type of selection will reduce the frequency of an undesired recessive phenotype, but it cannot eliminate it and produce a true-breeding deformity-free population.

For example, say a tilapia farmer has a population of Nile tilapia that contains both normally pigmented and red individuals, and decides to produce only red tilapia. If he tries to produce a true-breeding population of red Nile tilapia by culling all normally pigmented fish, he will discover that he will be unable to produce a red population that breeds true. The select red brood fish will produce both red and normally pigmented offspring.

This is because red is the dominant phenotype, and it is produced by the dominant R allele (RR and fir genotypes), while normal pigmentation is the recessive phenotype, and it is produced by the recessive r allele (rr genotype).

Because the homozygous dominant and the heterozygous genotypes produce identical red phenotypes, some of the select red brood fish (the heterozygotes) will have a copy of the undesired r allele. When two heterozygous red brood fish mate, they will produce some normally pigmented offspring. The same concept applies for phenotypes that are controlled by two or more genes. If the desired phenotype is produced by more than one genotype, and if at least one of the genes can be in either the homozygous or heterozygous state, selection cannot create a true-breeding population.

As was the case for phenotypes produced by a single gene, some of the select brood fish will carry a copy of the undesired recessive allele(s) and, when two heterozygotes mate, they will produce offspring with the undesired phenotype(s).

For example, if a carp farmer has a population of common carp with all four scale phenotypes and decides to produce a true-breeding population of scaled common carp by selection, he will discover that he cannot create a true-breeding population.

This is because the select scaled brood fish will have two genotypes—SS, nn or Ss, nn-and it is impossible to tell heterozygous scaled fish from homozygous scaled fish.

The farmer can eliminate the dominant N allele by culling all fish with the line and leather phenotypes (a dominant allele can be culled by a single act of selection); those individuals with two copies of the N, allele were culled by nature and died.

However, the farmer cannot eliminate the undesired s allele by culling the remaining phenotype-mirror. The select population of scaled brood fish will produce both mirror and scaled offspring when two select heterozygous (Ss, nn) scaled brood fish mate.

Progeny Testing: Since selection against (culling) a recessive phenotype cannot fix a dominant allele and produce a true-breeding population (when the mode of gene action is complete dominance), another type of breeding programme must be used to achieve this goal.

The only way an undesired recessive allele can be eliminated is by progeny testing. Progeny testing is a breeding programme that deciphers a parent's genotype by determining its offspring's phenotypes.

Once the genotypes of the dominant parents are deciphered, selection is used to save the homozygotes and cull the heterozygotes, which will eliminate all copies of the recessive allele, fix the dominant allele, and produce a true-breeding population.

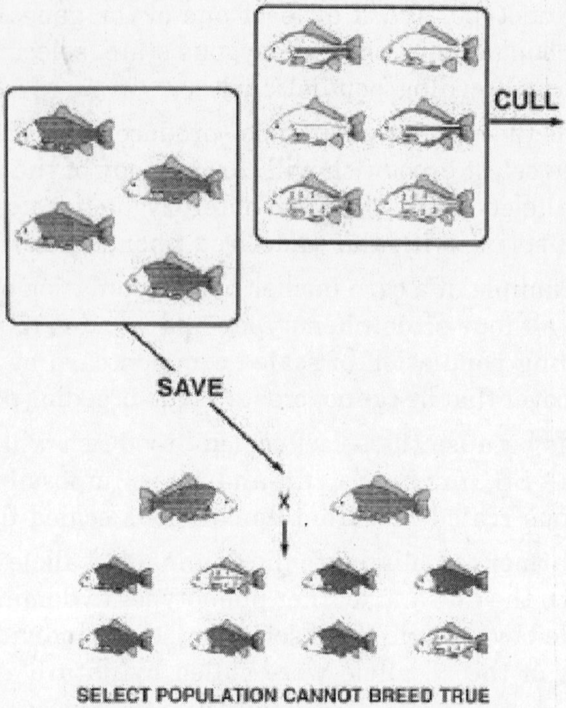

***Figure:** Selective breeding programme that is used in an unsuccessful attempt to produce a true-breeding population of scaled common carp. If all fish with the line, mirror, and leather phenotypes are culled, the select population will be composed of scaled fish. This act of selection will eliminate all copies of the undesired dominant N allele, because every heterozygote (line and leather phenotypes) will be culled (the homozygotes were culled by nature, since the NN genotype is lethal). Unfortunately, this selective breeding programme will not produce a true-breeding population because you cannot eliminate the undesired recessive s allele by culling the recessive phenotype (mirror). Because the scaled phenotype is produced by a dominant allele that exhibits complete dominance, both homozygous (SS, nn) and heterozygous (Ss, nn) scaled fish will be saved. Every heterozygous select scaled brood fish will carry a copy of the recessive s allele to the select population; consequently, selection cannot eliminate it. When select scaled brood fish are spawned, the mating of two heterozygotes will produce some offspring with the mirror phenotype (the mating that is depicted). As was the case with a phenotype produced by a single gene, if the desired phenotype is produced by more than one genotype and if one of the genotypes is a heterozygous genotype, selection cannot create a true-breeding population.*

This is accomplished by pairing and mating fish with the dominant phenotype to test fish (a fish whose genotype is known). Test fish are usually fish with the recessive phenotype, because such fish are homozygous recessive (experiments are not needed to determine a recessive phenotype's genotype). Because a test fish can produce gametes with only a recessive allele, the dominant fish is the parent that determines its offspring's phenotypes. If the dominant parent is homozygous, all offspring will have the dominant phenotype; if the dominant parent is a heterozygote, half the offspring will have the dominant phenotype and half will have the recessive phenotype.

The production of a single offspring with the recessive phenotype is enough evidence to state with certainty that the dominant parent is a heterozygote and must be culled. If no offspring with the recessive phenotype is detected in a random sample of at least 20 offspring, the dominant parent can be declared to be homozygous and kept for breeding purposes. Once a sufficient number of males and females are progeny tested and declared to be homozygous dominant, all other fish (including all offspring produced in the progeny tests) are culled. The population will now breed true and produce only offspring with the dominant phenotype.

For example, a tilapia farmer that has a population of red and normally pigmented Nile tilapia can create a true-breeding red population by using progeny testing to identify and keep red fish that are homozygous (RR) and to identify and cull red fish that are heterozygous (Rr). Before initiating the progeny testing programme, the farmer will cull all normally pigmented fish except for the ones that he will use as test fish.

The farmer progeny tests the red fish by pairing a red fish with a test fish (in this case, normally pigmented tilapia are used as test fish since they are homozygous recessive rr). Red parents that are homozygotes (RR) will produce only red offspring and will be saved; red parents that are heterozygotes (Rr) will produce equal numbers of red and normally pigmented offspring and will be culled. After one generation of progeny testing, the farmer will have a select population of red brood fish that will breed true and produce only red offspring, because every select red brood fish is homozygous RR.

The same principle applies if the desired phenotype is controlled by two or more genotypes and the phenotype is controlled by more than one gene. For example, if a carp farmer has a population of common carp with all four scale patterns and decides to produce a true-breeding population of scaled common carp, he will have to use progeny testing to accomplish his goal. Scaled fish are produced by both the SS, nn and Ss, nn genotypes.

In this case, fish with the mirror scale pattern are used as the test fish because they are homozygous recessive (ss, nn). Before initiating the progeny testing programme, the farmer will cull all fish with the line and leather phenotypes and most fish with the mirror phenotypes; he will keep a few fish with the mirror phenotype and use them as test fish.

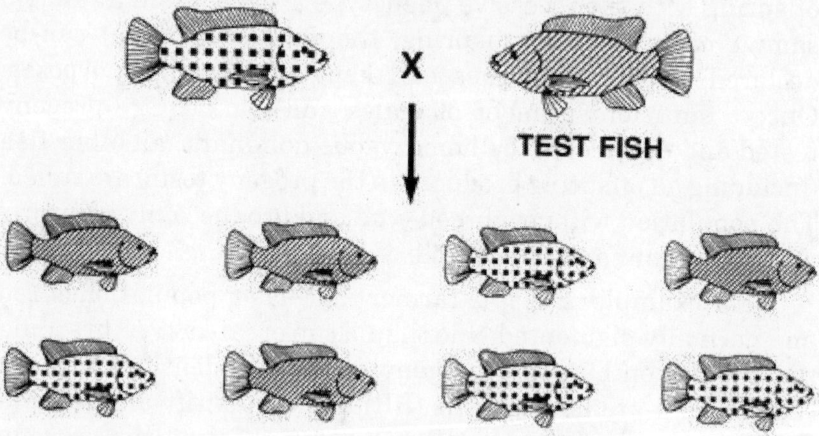

CULL RED FISH. IT IS A HETEROZYGOTE (*Rr*).

Progeny testing is done to identify and save scaled fish that are homozygous at the S locus (SS,nn) and to identify and cull scaled fish that are heterozygous at the S locus (Ss.nn). If the scaled parent is homozygous at the S locus (SS, nn), all offspring will have the scaled phenotype; if the scaled parent is a heterozygote (Ss, nn), half the offspring will have the scaled phenotype and half will have the mirror phenotype.

After one generation of progeny testing, the farmer will have a select population of scaled brood fish that are all SS, nn, and it will breed true and produce only scaled progeny.

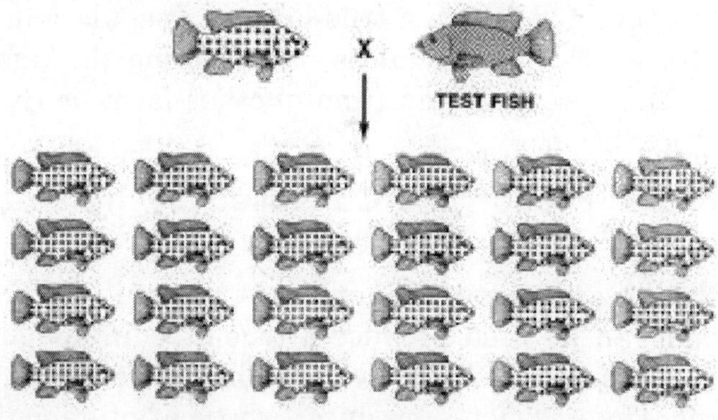

Figure: *Schematic diagram of the progeny testing programme that must be used to create a true-breeding population of red Nile tilapia. Progeny testing is used to identify and cull the heterozygous (Rr) red fish and to identify and save the homozygous (RR) red fish. A red fish's genotype is deciphered by mating it to a test fish (a normally pigmented fish [rr]), and by then determining the phenotypes of its offspring. In the top mating, the red fish is identified as a heterozygote (Rr) and is culled, because half of its offspring are normally pigmented. In the bottom mating, the red fish is considered to be a homozygote (RR) and is saved, because no normally pigmented fish was found in a random sample of a least 20 of its offspring. Progeny testing will create a population of select red brood fish that are all certified to be homozygous BE, and such a population will breed true and produce only red offspring. Pictorial representations of the phenotypes are the same as those used in Figure.*

When progeny testing, if the fish can be stripped, a single test fish can be used to progeny test a number of fish with the dominant phenotype. Additionally, it is not necessary to raise thousands of individuals from each family. A random sample of 100–200 fertilized eggs should be sufficient, if survival to the early fingerling stage is good. Consequently, progeny testing does not have to be expensive; it can be done in a small hatchery in aquaria, in small plastic swimming pools, or in hapas suspended in a single pond no larger than 0.04 ha. While the families are being raised, the dominant brood fish must be kept in isolated tanks or hapas so they can be saved or culled once the progeny are examined. Several fish can be kept in a tank or hapa if they are marked by fin clips. A single generation of progeny testing will create a true-breeding population, and it will not be needed again, unless fish carrying the undesired recessive allele are acquired from another farm or the allele arises by mutation.

Selection for Phenotypes Controlled by Heterozygous Genotypes

If the heterozygous phenotype is desired, no breeding programme can be used produce a true-breeding population. This is because fish with the desired phenotype are unable to breed true. Heterozygotes produce two types of gametes and produce them in equal numbers. Consequently, when they mate, they produce offspring with three phenotypes (selection for the heterozygous phenotype can only be done for phenotypes controlled by autosomal genes with incomplete dominance or with additive gene action). A farmer who culls the two homozygous phenotypes and creates a select population containing only the heterozygous phenotype, will quickly discover that the selective breeding programme is a failure, because only half the offspring produced by the select brood fish will have the desired phenotype.

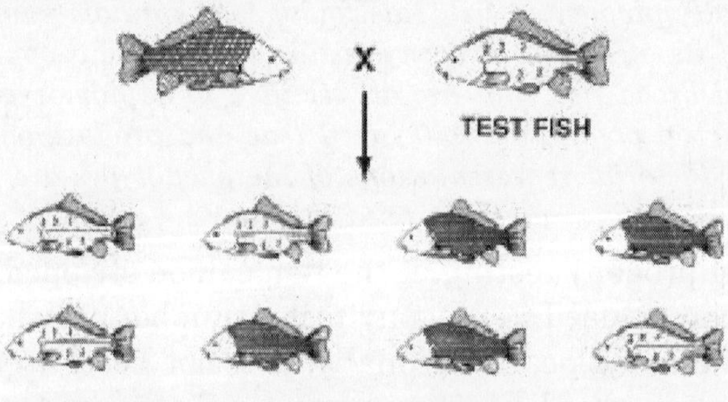

CULL SCALED FISH. IT IS A HETEROZYGOTE (*Ss*, *nn*).

For example, if a tilapia farmer has black, bronze, and gold phenotypes in his population of Mozambique tilapia and decides to produce a true-breeding population of bronze fish, he will find that it is impossible. If he conducts a selective breeding programme, he will cull the gold (the recessive phenotype) and the black (the dominant phenotype) fish and will save the bronze ones (the heterozygotes). When he breeds the select population of bronze brood fish, they will produce offspring with all three phenotypes. If the farmer wants to produce nothing but bronze fish, he must mate gold fish with black fish. This type of mating (the mating of the two homozygotes; it is illustrated in Figure) will produce 100% heterozygous offspring.

The only requirement for this breeding programme is that the farmer must be able to accurately sex his fish. If the sex of a single fish is mis-identified, gold or black offspring will be produced. This type of mating scheme requires the use of at least two ponds (or tanks) for offspring production: one will be stocked with gold females and black males; the other will be stocked with black females and gold males.

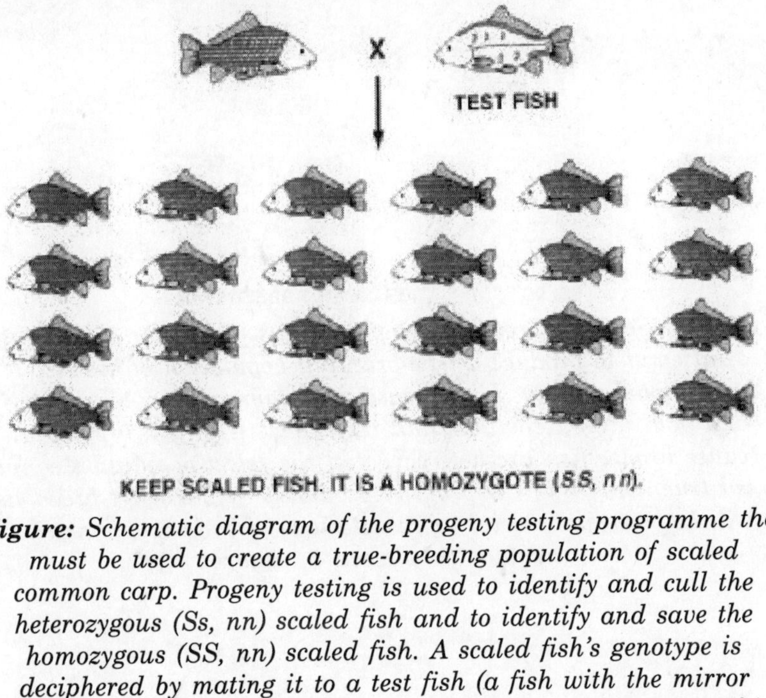

Figure: *Schematic diagram of the progeny testing programme that must be used to create a true-breeding population of scaled common carp. Progeny testing is used to identify and cull the heterozygous (Ss, nn) scaled fish and to identify and save the homozygous (SS, nn) scaled fish. A scaled fish's genotype is deciphered by mating it to a test fish (a fish with the mirror phenotype [ss, nn]), and by then determining the phenotypes of its offspring. In the top mating, the scaled fish is identified as a heterozygote (Ss, nn) and is culled, because half of its progeny have the mirror phenotype. In the bottom mating, the scaled fish is considered to be a homozygote (SS, nn) and is saved, because no fish with the mirror phenotype was found in a random sample of at least 20 of its offspring. Progeny testing will create a select population of scaled brood fish that are all certified to be homozygous SS, nn, and such a population will breed true and produce only scaled offspring.*

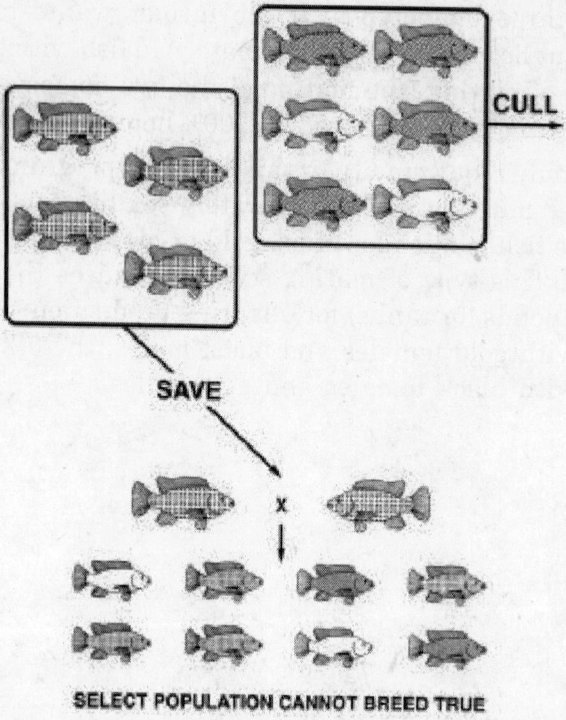

Figure: *Selective breeding programme used in an unsuccessful attempt to produce a true-breeding population of bronze Mozambique tilapia. In this case, both homozygotes-black (GG) and gold (ggl)-are culled, and the bronze (Gg) fish are saved. Because bronze fish are heterozygotes, the select population cannot breed true. Only half the progeny produced by the select brood fish will be bronze. Pictorial representations of body colour are the same as those used in Figure..*

Chapter 6

Selection for Quantitative Phenotypes

Quantitative phenotypes are the important production traits, such as length, weight, feed conversion, and eggs/kg female. When working with these traits, a farmer does not separate fish into descriptive non-overlapping categories as he does with qualitative phenotypes (for example, red vs normally pigmented). Instead, each trait is a single phenotype (for example, length), and the differences among individuals are determined by measuring the fish. The numerical value that is assigned to a fish when it is measured is an individual's phenotypic value.

Because an individual's phenotypic value is determined by measurement (for example, length in millimetres) rather than by descriptive category (for example, colour), selection is based on numerical values rather than appearance. For example, when selecting for a qualitative phenotype such as colour, farmers select fish which have a specific colour, and cull fish which have another; the decision to select or cull fish is based solely on descriptive category. When selecting for a quantitative phenotype, farmers select fish that meet or exceed a specific numerical value, and cull those that do not; the decision to select or cull fish is based solely on numerical values.

Selective breeding programmes that work with qualitative phenotypes are designed to create true-breeding populations. They are simple programmes that are designed to eliminate an undesired allele and fix a desired allele so the population breeds true and produces only the desired phenotype. This can usually be accomplished in a single generation (often by a single day of selection), and once this is accomplished, the farmer's goal has been achieved and the breeding programme is over.

In contrast, selective breeding programmes that are used to improve quantitative phenotypes are more sophisticated than those that are used to fix qualitative phenotypes. They require a higher level of technology and record keeping. Furthermore, they are open-ended in that they continue year after year, because it is not possible to produce true-breeding populations for these phenotypes. Instead, these programmes are designed to gradually improve the population's mean each generation. It is important to realise the fact that selective breeding programmes that work with quantitative phenotypes are never finished, and it is also important to strive for small but gradual improvements over many generations. It is possible to achieve huge improvements over the first one or two generations, but this can be accomplished only by severely reducing the size of the breeding population, which can result in inbreeding depression, and this could counteract much of the gains that were made by selection.

Another reason selective breeding programmes that are used to improve quantitative traits are open-ended is because if they are abandoned, much of the gain that was realised will be lost. It is difficult to predict how much gain will be lost if a programme is abandoned; that depends on the trait, the heritability, and the population. If the phenotype is controlled solely by additive genetic variance ($h = 1.0$), the mean will stay the same if the breeding programme is abandoned. However, no important production phenotype is controlled solely by additive genetic variance (Table 6). If a selective breeding programme is abandoned, much of the gain will be lost, because most of the gain due to dominance genetic variance, all of the gain due to epistatic genetic variance, and some of the gain due to genetic-environmental interactions will also be lost. Some of the gain due to additive genetic variance will remain, but the exact amount is difficult to predict.

If selective breeding greatly improved the population before the programme was abandoned, the mean might stabilize at a level recognisably greater than the original mean. However, if the gain was small, which is probably the case, the final mean might be only marginally greater than the original mean, and the difference might be obscured by yearly environmental effects.

Factors that must be Considered before Conducting Selective Breeding Programmes

Before initiating a selective breeding programme, a farmer must ask the following question: Is a selective breeding programme appropriate or necessary? The answer to this question depends on individual circumstances. The answer can even be different for neighbouring farmers.

The decision to conduct or not to conduct a selective breeding programme must be made at the individual level; it should never be made on a region-wide basis. There is a tendency by extension agents to try and get all farmers in their region to adopt new technologies. On an intellectual level, this makes sense because the extension agent knows that if every farmer adopts the new technologies, production will increase; farmers will get richer; and there will be more food.

The adoption of a selective breeding programme, like any aspect of technology, should be made only on a case-by-case basis. For the most part, the adoption of a selective breeding programme should be restricted to farmers who are capable of keeping and maintaining records, who are good managers, who adopt and maintain new technologies, and who are willing to incorporate some long-term planning into their management programmes. Finally, these farmers must be willing to spend the extra time that it takes to conduct a selective breeding programme and must also be willing to spend the money that it takes to conduct a programme.

Some farmers will never use nor should they be encouraged to use selection. These farmers are those that have small farms or that produce fish for only one or several families or perhaps for a small village. Farmers who do not have the expertise to keep good records or who do not have a sufficient number of ponds to devote to a breeding programme or who are unable or unwilling to spend money on a breeding programme should not attempt one or even be encouraged to attempt one. A selective breeding programme is designed to improve productivity over generations (a generation is the replacement of brood fish by their offspring). It is designed to increase growth rates and yields gradually over many years. If a farmer cannot incorporate long-term (1–10 years) planning into his fish farming programme, he will abandon his selective breeding programme within months and will then discourage other farmers from initiating a selective breeding programme by telling them that selective breeding programmes do not work.

Farmers who cannot keep good records because they cannot read or write or who do not know how to measure fish are not good candidates to become breeders and conduct a selective breeding programme. In general, these types of farmers will not be able to conduct a successful breeding programme. On the other hand, if they are reliable farmers who work hard and who accept and adopt new technologies, they can initiate a selective breeding programme if an extension agent is willing to work closely with them and assume the responsibility for keeping records and for conducting much of the labour involved in the breeding programme.

Finally, farmers who are poor managers should never conduct a selective breeding programme. Farmers who seldom check their ponds, who feed the fish or manure the ponds only when they are reminded by the extension agent, or who make little effort to increase yields should not conduct a selective breeding programme. Farmers who cannot or who are unable to properly manage a fish pond are not going to be able to properly manage a selective breeding programme.

The major goal of most selective breeding programmes that are going to be used to improve quantitative phenotypes is to improve growth rate. This will also increase yield.

There are two basic approaches to improving yields: One is to use a breeding programme to improve the fish that are being cultured. The other is to improve the way the fish are being raised. Breeding programmes are expensive, while improvements in management are often inexpensive.

Consequently, if increases in growth rate and yield can be easily gained by alterations in management techniques or by the adoption of new management techniques, farmers should use this approach first. This approach can include the proper use of fertilizers or lime, stocking fish at the most appropriate density, stocking larger fingerlings in grow-out ponds, or feeding fish a prepared ration.

The major reason why the first approach to increasing yields and growth rates is to use better management is that it is usually cheaper and the results are immediate. Yields can often double or triple if ponds are fertilized properly or if fish are fed. The increase in yield that occurs because of fertilization or feeding can be realised during the first growing season and during the first harvest. Yields will increase as a result of a selective breeding programme, but the increase will be gradual and it will be over many years. For this reason alone, the first approach to improving productivity should be to improve management skills and to improve the way the fish are grown.

Additionally, you do not want to initiate a selective breeding programme and then change the way the fish are cultured after the programme has begun. When you select for quantitative phenotypes, you select for them as they are expressed under a specific set of culture conditions. The fish that perform best under high stocking rates in manured ponds may not be the ones that perform best under lower stocking rates when the fish are fed a prepared ration. When you select for the fastest growing fish under one set of culture conditions you are selecting for fish that contain genes that enable the fish to perform best under those conditions. Consequently, if you change the way the fish are raised, you may have already culled the genes that perform well under the new management programme.

Thus, it is important to determine if the management programme that is being used to produce fish at a given farm is that which will be used in the future. Small changes, such as changing the stocking rate in grow-out ponds from 2,000/ha to 2,500/ha or increasing the mean size of fingerlings that are stocked from 10 g to 15 g, are probably not that important; but large changes, such as changing from manure as the sole source of nutrients to a high-quality pelleted ration, are major alterations in management. If major changes are planned or are likely, then selective breeding programmes should be postponed. Culture techniques in a region do not have to be standardized before selective breeding programmes are initiated.

Some farmers may want to culture fish using fertilizers rather than pelleted feed because they have abundant and cheap sources of manures; others may want to use pelleted rations to maximize yields. Selective breeding programmes can be designed for both types of farmers, if they are willing to devote the time and money to conduct the programme and if they can devote sufficient pond space for the programme.

When should a selective breeding programme be initiated? A programme should be initiated only when the farmer decides he is willing to devote the time and effort that will be needed to ensure its success. Farmers who do not want to devote the effort either in terms of manpower or money or who are likely to abandon the project after the initial excitement wears off should not start a selective breeding programme.

Some farmers may be interested in a selective breeding programme but say that they will start one only after they solve all other problems. This means that they will never conduct a selective breeding programme, because all other problems will never be solved. Every time one problem is solved, another roadblock to increased production appears.

A selective breeding programme should begin when a farmer is willing to spend the time and money needed to conduct one, because he is convinced that such an undertaking is a way of increasing both food production and profits, and because he also realises that these increased yields and profits are long-term goals. Finally, a farmer must also recognise that selective breeding programmes are open-ended, in that further improvements are always possible.

Before a farmer embarks on a selective breeding programme he must conduct an assessment to determine its usefulness; he needs to quantify his goals; he needs to know how the phenotype(s) will be measured; he needs to determine the type of selective breeding programme he will use; he needs to know how the results will be evaluated; and he must plan an annual evaluation and reassessment of the programme. A farmer can do this by himself or in consultation with extension agents.

The first step is to determine whether the programme is needed. Promoters often convince farmers that they need to make changes when none was needed. A salesman may have been impressed with the body shape of a fish that he saw in another province and feels that it would be a great idea if farmers would select for that body shape. If consumers truly desire that body shape and market surveys bear this out, then a farmer should consider a selective breeding programme to alter body shape. But if local consumers do not care and if there is no demand for the altered body shape, then the programme should not be initiated. Frivolous breeding programmes are a waste of money and manpower and may be counterproductive. The farmer must always consider what the consumer wants and what the consumer is willing to pay for when he initiates a selective breeding programme.

Another part of the assessment is to determine what traits should undergo selection. A farmer should prepare a wish list of all traits that he would like to improve. Once created, the list should then be pared to a handful of traits that make sense, in terms of the culture of the fish and its marketing. Some traits, like body weight, will be included in almost every selective breeding programme. Other traits that can make the initial cut are those that a farmer would like to improve, simply because they would make fish farming less risky or easier, such as increased disease resistance or increased catchability.

Some traits are totally unnecessary and should be eliminated immediately. A fish processing plant may want farmers to improve dressing percentage or to decrease visceral fat content because it will improve profits for the plant owners. But unless the plant owners are willing to pay more for fish with a better dressing percentage or with less fat, these traits should not be incorporated into a selective breeding programme.

Farmers should also eliminate some traits that are more easily improved by management. For example, it make little sense to use selective breeding to increase fecundity, because it is much easier and far less expensive to increase egg production by altering hatchery management. It also makes little sense because most fish are already highly fecund. Improving egg production from 4,000 eggs/kg female to 4,500 eggs/kg female is really not that important, and it is not necessary. If a farmer wants to produce more eggs, all he has to do is spawn more females or spawn larger females. Improving egg production via selective breeding is expensive, and an increase in egg production will not occur for at least one generation. On the other hand, improving egg production by altering hatchery management is inexpensive, and the increase is immediate.

Farmers should try to improve only the major phenotypes that improve profits-either because the fish grow faster, grow more efficiently, or survive better or because consumers are willing to pay more for them.

Hatchery managers who work at fingerling production centres might consider selecting for traits which could improve an industry or open new area to fish farming. For example: increasing salinity tolerance might enable fish farmers to grow fish in thousands of hectares of brackish water ponds; selection for earlier spawning date might enable farmers either to produce marketable sized fish in one growing season rather than two or to produce larger fish which command a premium price.

A second aspect of the assessment should be to decide whether the farmer is going to produce genetically improved fingerlings solely for his own use or whether he will become a fingerling farmer and produce genetically improved fish for a local fish farming industry. This decision determines how much money must be spent on the programme and the scope of the programme. If a farmer is going to produce genetically improved fish only for his own use, he needs to use only a portion of his farm for the project. On the other hand, a farmer who wants to produce genetically improved fingerlings for the local fish farming industry may need to use most, if not all, of his farm for this project.

One of the major purposes of reducing the wish list down to a few phenotypes is that the efficiency by which a trait can be improved is inversely related to the number of phenotypes that are incorporated into the selective breeding programme. Consequently, if one or two phenotypes are far more important than the others, the list should include only the one or two most important ones. The inclusion of traits that are not crucial will slow down the rate of improvement for the important ones. For this reason, most selective breeding programmes should try to improve only one or two phenotypes. Some programmes may include a third trait; few should incorporate more than that, and the only ones that should are those conducted by geneticists at research stations. Once a farmer has pared his wish list to the one or two phenotypes that he would like to improve, he needs to determine how they will be defined and measured. Goals that cannot be precisely defined and phenotypes which are difficult to measure should be dropped. For example, every farmer wants to improve disease resistance, but what is disease resistance? Some may define it as "fish that do not get sick," while others may define it as "fish that do not die." There is a great difference in these definitions.

If you define the trait as "does not get sick," what does that mean? Does this mean that the fish were infected but exhibited no clinical signs of the disease, or does it mean that they never were infected? If the fish are grown in a pond, how will you be able to tell which fish were infected and which were not? Even "does not die" is not that precise, because so many variables enter into health management. Fish can die from non-infectious sources during a disease outbreak or over the course of a growing season.

Because this trait is difficult to measure, it is a trait that should probably be avoided by most farmers. Furthermore, if resistance to a particular disease is heritable, it will be improved over time by what is called "domestication selection." Domestication is the process by which the farm and the farmer's management selects for or against genes as a natural consequence of farming. Consequently, disease resistance is a trait that should be removed from most farmer's lists. Some advances are being made in selection for disease resistance; however, these breeding programmes are complicated and should be conducted by scientists at research stations.

Other traits that must be pared are those that a farmer wants to improve but that are difficult to measure because of the way the fish are raised. Farmers that feed fish with a pelleted ration want to improve feed conversion because, if this trait can be improved, the fish will convert feed more efficiently, which means that yields and profits will soar. But how does one determine feed conversions for individual fish? This can be accomplished only by raising fish in individual tanks or hapas, which would be prohibitively expensive. A farmer could use family selection to improve this trait, if he were able to raise dozens of families in hapas. Such a programme would be expensive, and improving feed conversion is not as important as improving growth rate. If they are needed, the development of selective breeding programmes to improve feed conversion should be left to geneticists who work at research stations.

Even traits that every farmer wants to improve and that will be incorporated into a selective breeding programme need to be accurately defined and measured. Every farmer wants to improve growth, but how should it be defined and how will it be measured? This is important, because it will determine the rate of progress that can be made via selection. Growth rate is usually defined as weight or as length at a given age, so the first question that must be answered is: Should growth be defined as weight or as length?

It is easy to accurately measure hundreds of fish to the nearest millimetre, and the fish are minimally stressed during the process. However, length includes head length and tail length. Fish with an overly long tail could be selected over heavier fish, and consumers buy their fish by the kilogram, not by the centimetre. Weight more accurately reflects how fish are bought and sold, but it is difficult to accurately weigh hundreds of fish to the nearest 0.1 g. Unless the farmer has access to an electronic scale, the process can be very tedious, and it often stresses the fish. Furthermore, the weight recorded for each fish includes water in the fish's mouth, faeces and food in its digestive tract, and head and viscera weight, so it too is subject to error. Although it would be best to measure growth as weight, a large correlation exists between length and weight (the two are highly related-an increase in one means there is an increase in the other), so assessing growth by measuring length is a good approximation of assessing growth by measuring body weight. If a farmer wants to remove some of the error that can occur when growth is assessed by measuring length, he could take standard lengths instead of total length, which would negate the effect of tail length on selection.

If more than one person is going to measure fish, a standard method for measuring them must be determined and worked out in advance so that everyone will provide uniform measurements. Otherwise, the phenotypic values assigned to the fish (lengths) will be subject to human error. If only one person will measure the fish, the only requirement is that he measure all fish using the same technique.

If weight is improved by selection for length, weight is said to have been improved by "indirect selection." This occurs because the two phenotypes are positively correlated (related genetically), so improving one will automatically improve the other. Because of this, selection to improve growth rate can often improve other traits, such as survival, disease resistance, and feed conversion.

Breeders often exploit genetic correlations and use indirect selection to improve phenotypes that are difficult or expensive to measure by selecting for more easily measured phenotypes. For example, harvest weight in rainbow trout has been improved via indirect selection by selecting for fingerling weight; selective breeding programmes are being conducted to improve disease resistance via indirect selection by selecting for stress response or for cell membrane strength.

Another question that must be answered before growth rate can be improved is: At what age will growth be assessed? Fish growth is not constant. If it were, the age at which the measurement was taken would be relatively unimportant. It makes sense to measure fish when they are normally moved or harvested. For example, fish can be measured when they are transferred from fingerling ponds to grow-out ponds and when they are harvested for market. This is usually defined as measurement at a specific age. After a number of generations of selection, it might be defined as the time needed to reach market size. If fish are measured and selected at two ages instead of one, growth rate will be improved more rapidly.

The ability to precisely define the goal and to accurately measure the phenotype is crucial. If a farmer cannot perform these tasks, he should not conduct a selective breeding programme. If he conducts a selective breeding programme using imprecisely defined goals and measures the phenotype inaccurately, his breeding programme can decrease growth rate and may also prevent future improvements by culling the truly valuable alleles.

For example, selection for growth rate in some populations of Indian carps produced slower-growing fish because growth rate was imprecisely defined and measured inaccurately. These carps were cultured using a production system called "multiple batch" production. When this form of management is used, a pond is seined many times over a several-year period to continually remove market-sized fish. Fingerlings are often stocked in the pond to replace fish that are removed to ensure that fish can be continually harvested. After several years of continual production, the pond is drained because large fish that have escaped the repeated seinings adversely affect production.

When these carp ponds were drained, farmers and hatchery personnel saved the largest fish, because they defined growth rate as body size at draining, and measured it only when a multiple batch production pond was drained. Additionally, they saved these fish because they would produce more eggs than smaller brood fish.

Unfortunately, their definition of growth rate and the way the trait was measured (size when the ponds were drained) produced a selective breeding programme that reduced growth rate, because the select fish were actually the oldest fish and were are also slow-growing fish. They were the largest fish when the ponds were drained simply because they were experts at escaping a seine. Not only did this selective breeding programme produce shower-growing fish, it also selected for fish that were escape artists.

Once a farmer has quantified his goals and has determined how the phenotypes will be measured, he needs to determine what type of selective breeding programme he will use to achieve his goal. There are two basic types of selective breeding programmes: individual selection (also called "mass selection") and family selection. In individual selection, family relationships are ignored, and comparisons are made among individuals. Individual fish are ranked in order, and the best become the select brood fish. In family selection, family relationships are of paramount important, and either family means are compared and ranked and whole families are either saved or culled, or individual rankings are made within each family and selection occurs independently within each family. The most efficient breeding programme is determined by the phenotype and how it will be measured, the biology of the fish, and by the trait's heritability. Individual and family selection will be discussed in greater detail later in this chapter.

A farmer needs to outline the plan he will use to achieve his goal. This includes the number of fish that will be spawned, the number of fish in each family that will be raised, the number of ponds that will be stocked, the stocking density, and the management programme that will be used to culture the fish.

Selective breeding programmes are expensive, so it is important to be able to accurately assess the effects of selection. The best way to determine this is to compare the phenotype's mean over time. To do this properly, the mean of the select population must be compared to that from an unselected control population. The reason the select population must be compared to a control population is that the mean is determined by both the population's genome (which you are trying to improve via selection) and by the environment. Techniques that are needed to create a control population and the way it is used to assess the results of selection will be discussed in a later section of this chapter.

Finally, a farmer must review his selective breeding programme at least once a year. He must evaluate how it is progressing and determine if there are ways to improve the programme. Conversely, he may decide to end the programme if the results indicate that the programme is not working or if the results are not worth the effort. Farmers often want to end one selective breeding programme and start another when they get frustrated, but this type of decision should be discouraged. If a farmer changes programmes every year or two, all he will do is spend money and accomplish nothing. On the other hand, if the programme is not working, it would be foolish to persist and spend money needlessly.

Environmental Effects that must be Controlled

Environmental variables have a significant influence on the production of quantitative phenotypes. Even when a phenotype has a large heritability, the environmental component of phenotypic variance can be larger than the heritable component. If the environmental influence on phenotypic variance were the same for all fish, it would be of minimal concern during selection, as long as there is sufficient heritable variance to enable selection to work.

Unfortunately, many sources of environmental variance have a differential effect on individuals or families and, when selection is used, a farmer must control these environmental variables or they will give some individuals or families an advantage. If this occurs, the select brood fish may be better because of an environmental advantage, not because they are genetically superior. If this happens, selection will not work.

For example, older fish are usually the largest fish in a population because they have had more growing days. It has been shown that an age difference as small as 24 hours can give the older fish a size advantage that they never relinquish, which means that selection will choose the older fish, not the genetically superior fish; consequently, if age-related size differences are not controlled, a selective breeding programme that is trying to improve growth rate would actually be trying to improve it by exploiting environmental variance (age) rather than additive genetic variance, and the programme would be a failure.

Thus, if it is possible, all fish should be spawned on the same day. If this is not possible, then the population should be split into cohorts of equal-aged fish, and selection can occur independently in each cohort. After selection, the select brood fish from each cohort can be mixed.

Other factors that need to be controlled are female age and female size and management differences among the fingerling or the grow-out ponds. For example, if each family is stocked in a single pond, individual selection will be unable to differentiate the fish that are largest because they are genetically superior from those that are largest because they were grown at the lowest density or had the most food.

If a farmer does not control these variables, the largest fish may all come from one family, and they may be largest because of environmental variance. If this is the case, no progress can be made by selection. Even seemingly insignificant environmental variables such as feed particle size can be a major sources of phenotypic variance. Studies with common carp have shown that if feed particles are improperly sized, some individuals will grow significantly faster than the rest of the population; the size advantage that these fish have is not exploitable in a breeding programme, because it is an environmental advantage, not a heritable one.

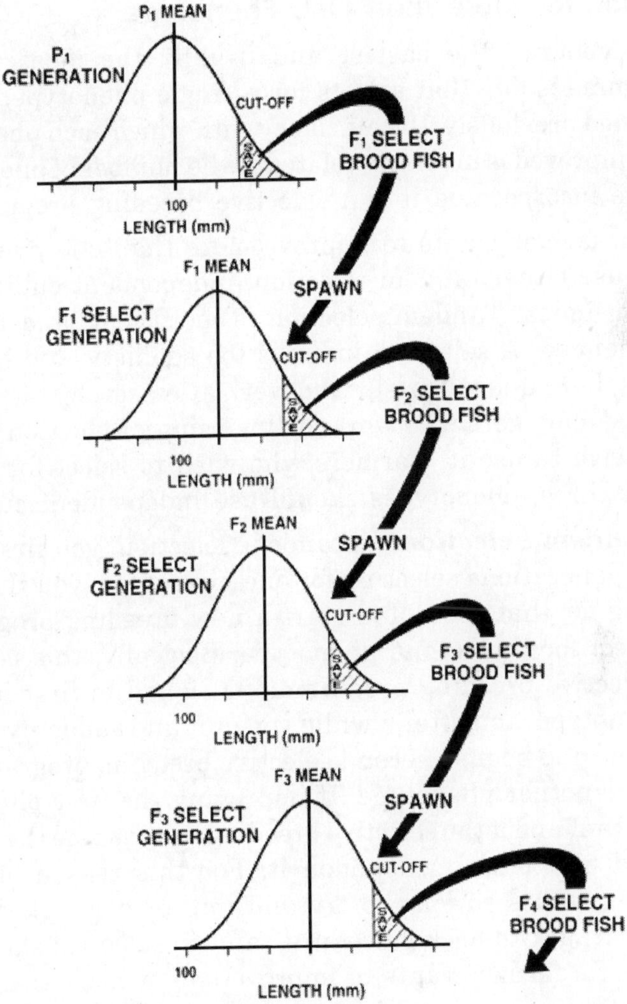

Figure: *Schematic diagram of individual selection to improve growth rate by selecting for increased length. The figure shows four generations of selection and shows how length can be improved over time.*

When conducting a selective breeding programme, a farmer must control all environmental factors that can give individuals or families an advantage. If these variables are not controlled, selection will either be less effective than it is for others who are conducting similar programmes or it will not work. Since selective breeding programmes require considerable effort and time, the inability to control these variables will be costly.

Selection for more than One Phenotype

In general, the easiest and usually the most efficient programme is one that selects for a single phenotype. As was mentioned previously, the efficiency with which each phenotype' can be improved is inversely related to the number of phenotypes that are incorporated into a selective breeding programme.

If a farmer wants to improve more than one phenotype, he can use either tandem selection, independent culling, or a selection index. Tandem selection is inefficient, and it is not recommended. A selection index is too sophisticated for most farmers, but aquaculturists who work at research stations and hatchery managers who work at fingerling production centres might wish to use it. Farmers who wish to select for two, or at most three, phenotypes should use independent culling.

Tandem Selection: In tandem selection, you first spend several generations selecting for one phenotype; you then stop selecting for that trait, and start a new breeding programme and select for the second phenotype. Basically, this is a two-step selective breeding programme designed to first improve one phenotype, and after a while you quit and suddenly change directions and begin a second selective breeding programme to improve another phenotype. If improving the first phenotype is important and if the selective breeding programme is working, it would be foolish to abandon it. For this reason and also because it takes so long to try and improve two phenotypes using this back-to-back procedure, tandem selection should not be used if a farmer wants to improve two or more traits. It is very inefficient.

Independent Culling: Independent culling is probably the easiest selective breeding programme that a farmer can use to improve two or three phenotypes. Independent culling is a breeding programme where a farmer selects simultaneously for two or more phenotypes by establishing cut-off values for each phenotype independent of the other(s), and each fish must reach or exceed all cut-off values in order to be saved. A fish which fails one cut-off value is culled. This type of breeding programme is illustrated in Figure.

For example, say a farmer wants to improve both growth rate, by selecting for length, and body conformation, by selecting for body depth at the first dorsal fin spine.

The procedure that is used to establish the cut-off values is the same as that outlined earlier. The farmer will either establish pre-determined phenotypic values or will select the top 10% or top 20%, etc.

If he establishes pre-determined percentiles, he determines the cut-off values that correspond to these percentiles by measuring a random sample of 100–200 fish and by discovering what those values are, as was illustrated in Figure.

In this example, the farmer has established cut-off values of 243 mm for length and 130 mm for body depth at the first dorsal fin spine. Once both cut-off values are established, the farmer measures each fish, and only those that meet or exceed both cut-off values will be saved. In this breeding programme, the decision to save or to cull these six fish will be:

Fish	*Length (mm)*	*Depth (mm)*	*Save/Cull*
1	254	129	cull
2	243	130	save
3	241	138	cull
4	240	120	cull
5	280	129	cull
6	250	131	save

Although independent culling is more efficient than tandem selection and it is easy to conduct, this type of breeding programme has two liabilities. The first is that when you select simultaneously and independently for two phenotypes, you could end up with very few select fish. For example, if you select for two phenotypes and the cut-off value is the top 10% for both traits, only 1% of the population will be saved ($0.1 \times 0.1 = 0.01$). If you add a third phenotype and the cut-off value for it is also at that level, only one fish out of every thousand will be saved. The severity of this type of selection can be seen in Figure. The severe restriction in the number of fish that are able to meet or exceed all cut-off values can lead to two problems:

If this occurs, you may not be able to produce enough select fingerlings for grow-out the following season. Even worse, high levels of inbreeding could be produced after 1–2 generations, which could negate much of the gains made by selection.

The problem created by independent cut-off values can be solved by deciding in advance how severe selection will be; that is, what percentage of fish will be selected.

Once this is determined, you back-calculate the cut-off percentages that are needed to produce this value. For example, if you want to retain 15% of the population and will select for two equally important phenotypes, simply take the square root of 0.15, and that tells you to select the top 39% for both phenotypes (0.39 × 0.39 = 0.15).

If you want to put more emphasis on one phenotype, simply establish a higher cut-off value (in terms of percentiles) for that trait and then adjust the cut-off value for the second downwards.

For example, if you want to save 10% of the population, you can have the cut-off values at the 25th and 40th percentiles for two phenotypes (0.4 × 0.25 = 0.1).

A second draw-back that occurs as a result of independent culling is the fact that you must cull fish that are outstanding in one phenotype.

For example, fish number 5 in the example given above was culled despite the fact that it far exceeded the length cut-off value. This problem can be circumvented by using modified independent culling. In modified independent culling, fish that meet or exceed both cut-off values are saved, and fish that exceed one cut-off value but that are just under the cut-off value for the second phenotype are also saved. Because of the ease with which they can be conducted and because the record keeping is not that cumbersome, independent culling or modified independent culling are the selective breeding programmes that most farmers should use if they want to simultaneously improve two or perhaps three phenotypes.

Selection Index: A selection index is the most efficient selective breeding programme that can be used to improve two or more phenotypes. Unfortunately, it is also more sophisticated and more expensive than independent culling.

This type of selective breeding programme should only be used by aquaculturists who work at research stations or by hatchery managers who work at fingerling production centres. Farmers and most aquaculturists should use independent culling if they want to improve two phenotypes. A selection index is really only needed if three or more phenotypes will be improved simultaneously.

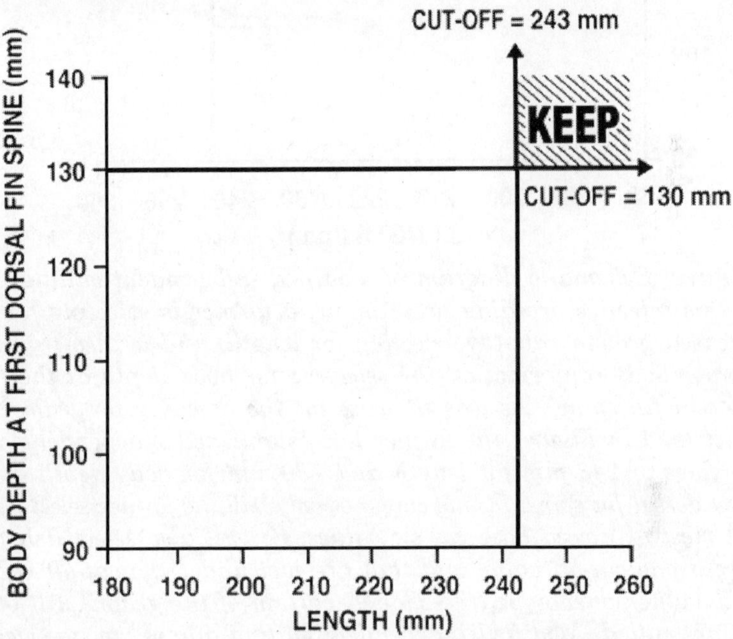

Figure: *Schematic diagram of independent culling. In this selective breeding programme, a farmer is selecting for increased growth rate (by selecting for length) and is also trying to improve body conformation (by selecting for body depth at the first dorsal fin spine). He has established independent cut-off values of 243 mm for length and 130 mm for body depth at the first dorsal fin spine. Fish which meet or exceed both cut-off values are saved and become the select brood fish; all other fish are culled. The graph illustrates the relative percentages of fish that will be saved (shaded portion) and culled (unshaded portion) with this type of selective breeding programme.*

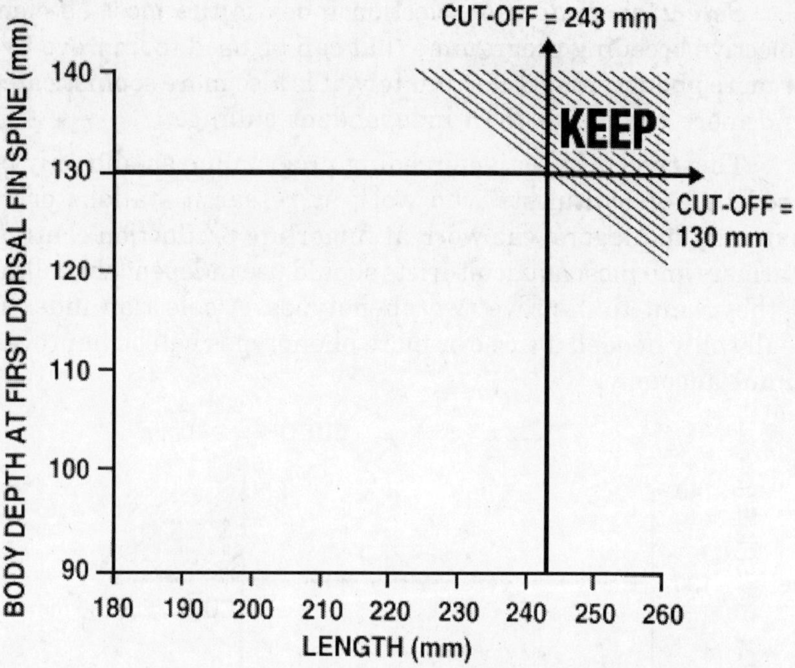

Figure: *Schematic diagram of modified independent culling. In this selective breeding programme, a farmer is selecting for increased growth rate (by selecting for length) and is also trying to improve body conformation (by selecting for body depth at the first dorsal fin spine). As was the case for the breeding programme illustrated in Figure, the farmer has established independent cut-off values of 243 mm for length and 130 mm for body depth at the first dorsal fin spine. In this case, not only will the farmer select fish that meet or exceed both cut-off values, he will also select fish that exceed one cut-off value and that are just under the cut-off value for the other phenotype (the shaded portion of the graph). All other fish are culled. Modified independent culling allows you to select a greater number of fish, and individuals which have a truly outstanding phenotypic value for one trait do not have to be culled.*

A selection index is a selective breeding programme where two or more phenotypes are simultaneously evaluated and an overall numerical value is assigned to each fish. The fish are then ranked by these numerical values. The rankings are established so that a fish that is at the 50th percentile for every phenotype receives an I value of 100.0. Consequently, fish that have I values > 100 are considered to be above average, while those with I values <100 are considered to be below average.

The ultimate goal of every large-scale sophisticated selective breeding programme is to develop a selection index. A selection index requires heritabilities, genetic correlations among the phenotypes, and the economic importance of the phenotypes; unfortunately, this information is not known for most cultured populations of food fish. The need for this information can be circumvented by modifying the selection index and by creating what are called "importance factors" for each phenotype.

The techniques that are used to create a selection index using importance factors are outlined in Table 11. The index in Table 11 is designed to improve three phenotypes: length at 12 months, body depth at the first dorsal fin spine, and dressing percentage (as approximated by the ratio: trunk length [length from the back of the head to the end of the caudal peduncle]/ total length). Once an I value is computed for each fish, the fish are ranked by I values and the top 10% or top 15% or top 20% are selected, while the remainder are culled. A selective breeding programme that uses the selection index derived in Table 11 is outlined in Figure.

This type of selective breeding programme requires far more labour than independent culling. In independent culling, a farmer assigns a cut-off value for each phenotype (usually based on a sample of 100–200 fish) which will enable him to quickly save or cull each fish. In a selection index, all phenotypes must be measured for each fish, and the fish must be given an individual tag or mark and placed in a tank or hapa until all fish are measured. After all fish are measured, the fish are once again identified and either saved or culled, based on their I values. Not only is this process labour intensive, but the extra handling stresses the fish. For all these reasons, if a farmer wants to improve two or perhaps three phenotypes, he should use either independent culling or modified independent culling.

Simple Selective Breeding Programmes to Improve Growth Rate and Other Quantitative Phenotypes

Even though selective breeding programmes can be designed to improve all sorts of production phenotypes, the single most important phenotype is growth rate. Improving growth rate will decrease the time it takes to grow a fish to market, which means a farmer can produce more crops in a given time period, which means he will make more money. It also means yields will increase. This will increase production efficiency, increase food production, and also increase a farmer's income. As an added bonus, increasing growth rate can improve other production phenotypes via indirect selection. Some studies have shown that faster-growing fish also have more efficient feed conversions and seem to be more disease resistant.

Selection can be used to improve other quantitative phenotypes, if improving them will improve production efficiencies or profits. In most cases, as important as these phenotypes are, they are not as important as growth rate. Occasionally, a breeding programme other than selection is needed to solve the most important goal. For example: in tilapia farming, controlling reproduction in grow-out ponds is the most important goal, and this is accomplished by interspecific hybridization and/or sex reversal; in grass carp farming in the U.S., producing sterile fish is the most important goal, and this is achieved by chromosomal manipulation.

The object of this chapter is to outline simple, relatively inexpensive selective breeding programmes that can be conducted by farmers on farms with about 2 ha of ponds and to provide examples of the types of data that must be collected, as well as data tables that can be used to record these data. This chapter will not discuss selective breeding programmes that use tandem selection; a selection index; or those that combine selection with crossbreeding, inbreeding, or some facet of biotechnology such as chromosomal manipulation.

When describing these simple selective breeding programmes, information will be provided on the number of ponds that will be needed in order to grow the fish that will be evaluated. This number will not include the number of ponds that will be needed to hold or to spawn brood stock. Additionally, this chapter will not discuss ancillary facilities that are needed, such as holding tanks, hatchery buildings, counting tables, etc.

The main goal for the selective breeding programmes that are outlined in this chapter will be to improve growth rate by selecting for length. As was described in Chapter 4, growth can be improved by selecting for either length or weight. Because it is far easier to accurately measure hundreds of fish than it is to accurately weigh them, and also because most farmers do not have access to accurate balances, most farmers should select for length.

Examples of selective breeding programmes that are designed to improve two phenotypes will also be outlined. In these examples, selection to improve growth rate is still the primary goal, but a second phenotype is added.

The selective breeding programmes outlined in this chapter are designed to be conducted by a farmer who is creating genetically improved stock for his farm. If a farmer wants to embark on a breeding programme so that he can sell genetically improved fingerlings to support a local or regional industry, the size and costs of the projects that are outlined must be increased: more ponds must be built; more cohorts must be created; and more select brood fish must be produced. In fact, if this is a farmer's goal, it is likely that he will convert his entire farm to this enterprise, and he will no longer produce food fish.

One of the things a farmer must do if he is to run a successful selective breeding programme is to determine how many select brood fish must be saved. If too few are saved, a farmer will be unable to produce enough select fish for future grow-out. If this happens, he will have to spawn unselected fish, which will negate much of his efforts.

An additional problem is that if a farmer saves too few fish, inbreeding can build to levels which will result in inbreeding depression. If this occurs, much of the improvements created by selective breeding would be used simply to counteract inbreeding depression.

A farmer should spawn at least 25 males and 25 females every generation to minimize inbreeding depression. This number is not carved in stone, but it can be used as a general rule. This means that a farmer should save a minimum of 100–200 brood fish. The reason why a farmer needs to save this many select brood fish is some will die before they can be spawned, and spawning success is seldom 100%. These guidelines also apply for a control population. A second factor that determines how many brood fish a farmer needs is the size of his operation; that is, how many fingerlings are needed for the grow-out ponds. If a farmer saves too few brood fish, he will not be able to stock his grow-out ponds. Saving enough select brood fish can be a problem on large fish farms, but it is seldom a problem on medium-sized (2 ha) fish farms, especially if a farmer saves at least 100–200 select brood fish.

The selective breeding programmes described in this chapter are bare-boned skeletal outlines that are presented to demonstrate that relatively simple and relatively inexpensive programmes can be conducted by farmers and that, if the programmes are conducted properly, they can be integrated into everyday farming practices.

The programmes outlined in this chapter are not absolute. They can be and should be modified in order to customize the programme for a particular species, for a farmer's income, and for his farm.

The breeding programmes that are outlined in this chapter are for species that do not exhibit sexual dimorphism. This means a single cut-off value can be used during selection.

If the species that a farmer grows exhibits sexual dimorphism a slight modification of the programmes that are outlined will be necessary. Sexual dimorphism can increase the effort needed to conduct selection, because once it occurs, the fish must be sexed and selection must occur independently in each sex. The age at which sexual dimorphism occurs will determine if a single cut-off value can be used or if separate cut-off values will be needed for males and females. In some species, sexual dimorphism occurs when the fish are small and separate cut-off values are needed at both the fingerling and food fish phases of selection. If sexual dimorphism does not occur until after the fingerling stage, a single cut-off value can be used at the fingerling stage, while separate ones will be needed when food fish are harvested. If sexual dimorphism occurs after the age at which food fish are harvested, a single cut-off value can be used during both phases of selection.

The selective breeding programmes outlined in this chapter do not describe the creation and use of a control population to assess the results of selection. This was described in Chapter 4.

Individual Selection

If possible, individual selection should be used to improve growth rate. Individual selection is easier and less expensive than family selection, because it can be done in only one or two ponds and fewer fish need to measured.

Selection for Growth Rate

Growth rate is the most important production phenotype. The following programmes outline how individual selection can be used to improve only this trait. When conducting individual selection, if a farmer can synchronize spawning, selection will be relatively simple. If the species spawns asynchronously, care must be taken to ensure that age-related size differences do not confound and obscure genetically-produced size differences. If the selective breeding programme is not modified to accommodate asynchronous spawning behaviour it will be impossible to identify fish which are genetically superior from those which are environmentally superior (older).

Synchronous Spawning: If a farmer can synchronize spawning and can produce at least 25 families on a single day (or at most, over a 48-hour period), he can conduct a simple, inexpensive selective breeding programme in only one or two ponds. In addition, an extra pond will be needed for the select brood fish, once they have been saved.

To initiate the selective breeding programme, the fish should be spawned using normal management techniques. If possible, egg masses should be collected and incubated. If the eggs are usually incubated by the females in the ponds, they should be closely monitored, and fry should be collected as soon as they hatch or begin to swim away from each female.

If possible, family size should be equalized before fish are stocked. This will prevent one family from skewing the results of selection; additionally, it will help minimize inbreeding. Because of this, families should be isolated until they are equalized. In addition, if families are isolated until they are stocked, the complete mortality of one or more families will be noticeable and can be recorded. If the fish are traditionally raised in a two-phase process (phase one is stocking fry and raising them to fingerlings; phase two is stocking fingerlings and raising them to food fish), selection can be done when fingerlings are harvested and when fish are harvested for market; if this is the case, selection will require two ponds. If production is only a single-phase process (no fingerling phase; fry are stocked and raised to food fish), selection will occur when fish are harvested for market, and selection will require only one pond.

If fish are raised in a two-phase process, fry should be stocked in a single pond and fingerlings should be produced using normal production techniques. Just prior to harvest, a random sample of 100–200 fingerlings should be measured to the nearest millimetre in order to determine the phenotypic value that corresponds to the desired cut-off percentage, as was outlined in Figure. At fingerling harvest, the top 35–50% should be saved and stocked in the food fish production pond. The culled fingerlings can be grown for food or sold. If they are grown for food, no fish from this population should be saved and spawned.

The select fingerlings should be raised using normal production management. Prior to harvest, a random sample of 100–200 fish should be measured to the nearest millimetre to determine the phenotypic value that corresponds to the desired cut-off percentage. At harvest, the top 10–20% of the fish should be saved, and these fish will become the select brood fish. Culled fish can be eaten or sold.

If fish are raised in a single-phase process, a random sample of 100–200 fish should be obtained to create the F_1 control brood fish when fish are harvested. The population of F_1 control brood fish should be created prior to selection. If fish are raised in a two-phase process and if selection will occur at both the fingerling and food fish harvests, the F_1 control brood fish need to be saved at the end of the fingerling phase. If the F_1 control brood fish are saved just prior to the second act of selection during a two-phase selection process, the control will not be a true control population, since some selection will have occurred prior to its creation.

If fish are raised in a single-phase process, fry should be stocked in a single pond, and fish should be produced using normal production techniques. Selection at harvest is done as described above. These simple selective breeding programmes are outlined in Figures. The cut-off percentages were mentioned as ranges, not as specific values (35–50% for fingerlings and 10–20% for food fish). This is because the exact cut-off values are not that critical. The intensities of selection that are used are individual decisions. A farmer can increase the rate of gain by using higher cut-off percentages (saving a smaller percentage). But rate of gain must be balanced against inbreeding-related problems and with the ability to produce enough fish in the next generation. Most selective breeding programmes conducted on medium-sized farms will encounter few problems and will also be able to achieve desired results if cut-off values in these ranges are used. The pond(s) used in this project should be 0.04–0.4 ha. Pond size is determined by the stocking rate that will be used to grow the fish and by the intensity of selection. Once a farmer decides what his cut-off percentages will be and how many select brood fish will be saved, he can determine how big the ponds must be. This will tell him the percent of the farm that will be devoted to the breeding programme. The procedure used to determine pond size is outlined in Table. The values derived in Table are valid only for the assumptions that were given in Table.

If ponds of approximately the correct size already exist, a farmer does not have to destroy them and custom-build new ponds in order to conduct a selective breeding programme. A farmer who uses existing ponds should adjust stocking rates and/or cut-off values to achieve the desired results. The formulae presented in Table 14 can be used to generate this information.

The information generated in Table 14 can also be used to determine how many fish are needed from each family. For example, if a farmer needs to stock 8,892 fry and if he has produced 25 families, he needs 355.7 fry/family (8,892/25). Because you cannot have 0.7 fry, he needs 356 fry/family. Because the number is rounded up, the farmer will stock 8,900 fry. If he maintains the stocking rate at 200,000 fry/ha, the size of his fingerling pond would change only by 0.001 ha to accommodate the 8 additional fry, so a larger pond would not be needed (pond size had already been rounded to the nearest thousandth). The rounding up of family size also means that the initial culling will produce an additional 2 select fingerlings. This will have little effect on stocking rate in the food fish pond, nor will it require a larger food fish pond. If the farmer does not want to alter stocking rate one iota from his planned rate, he can randomly cull 2 select fingerlings once that population has been created.

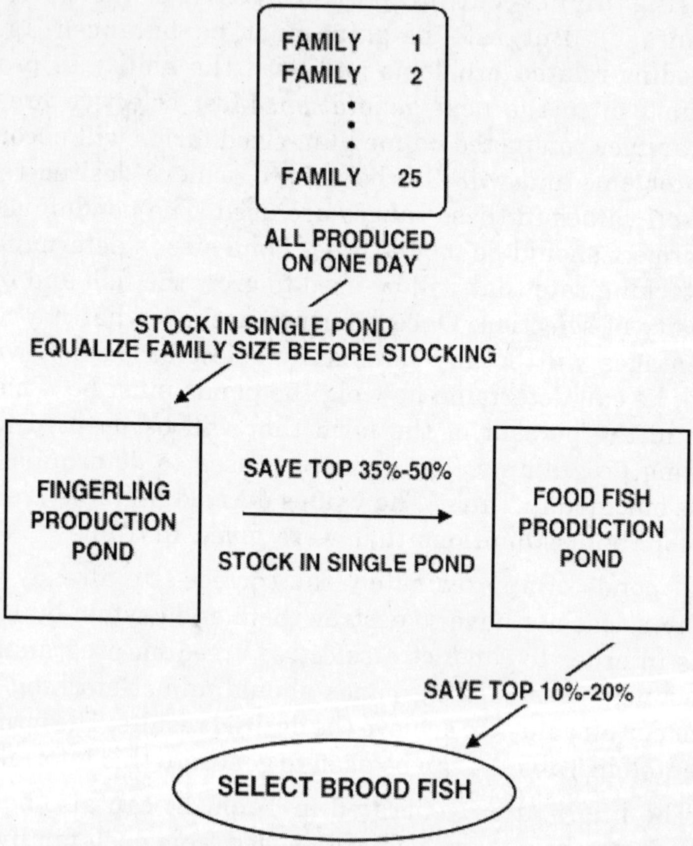

Figure: *Schematic diagram of a simple and inexpensive selective breeding programme to improve growth rate by selection at two ages-at fingerling harvest and at food fish harvest. This breeding programme can be conducted in two ponds. Fish that are culled at the fingerling stage can be sold or grown for food. Fish that are culled at harvest can be eaten; sold as food; or some can be retained and used as brood fish to produce the production fish (fish raised and sold as food), if the select fish cannot produce enough offspring for both the selective breeding programme and for the production ponds.*

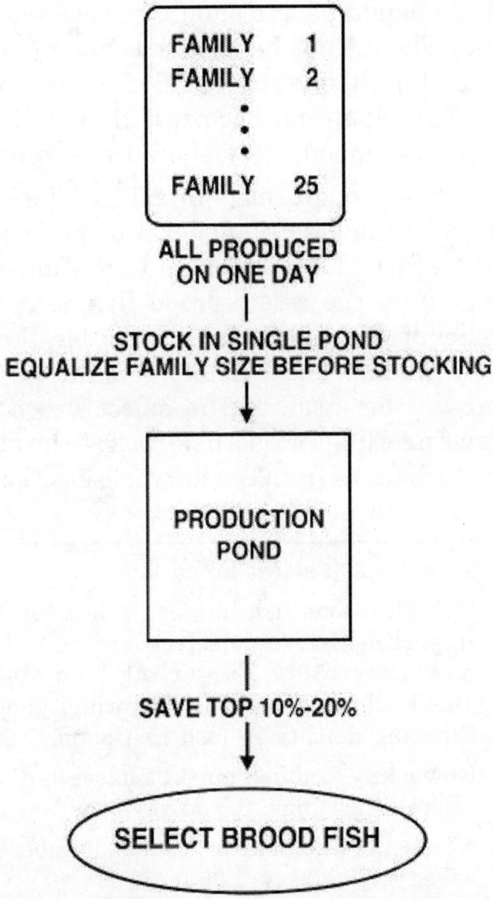

Figure: *Schematic diagram of the least expensive selective breeding programme that can be designed to improve growth rate. In this breeding programme, fry are stocked in a pond, and the fish are not harvested until they are ready for market. Selection occurs when the pond is drained and the fish are harvested. This breeding programme can be done in one pond. The fate of the culled fish was described in Figure.*

Brood fish that are saved should be stocked in a select brood stock pond. No other fish should be stocked with the select brood fish. When mature, these fish will be spawned to produce the F_1 select generation. No selection for secondary sexual traits and no selection for other criteria should be done with the brood fish prior to the mating season. The only goal of this selective breeding programme was to improve growth rate by selecting for length, and the select brood fish were selected for that phenotype and should be selected only for that phenotype; they should not be selected for any other reason. The select brood fish should be allowed to mate among themselves in a random manner or, if they will be paired in pens or manually stripped, they should be randomly paired.

If a farmer has enough select brood fish, he can use these fish to produce both the F_1 select generation and the production fish (fish that the farmer will stock and grow for market in his production ponds). If the select brood fish cannot produce a sufficient number of offspring for both purposes, their fry should first be used to produce the F_1 select generation; any surplus fish can be grown for food. If the select brood fish cannot produce enough fingerlings for both purposes, the farmer should use other brood fish (culls from the harvest selection) to produce fry for the production ponds.

Goal:	To have 200 select brood fish
Given:	Cut-off at food fish harvest: select top 10%Cut-off at fingerling harvest: select top 50% Fry-fingerling mortality: 50% Fingerling-food fish mortality: 10%Stocking density in fingerling pond: 200,000/ha Stocking density in food fish pond: 7,000/ha
Step 1.	How many food fish must be harvested to produce 200 select fingerlings, if you save the top 10%? Number of food fish harvested = number saved/percent saved = 200/0.1 = 2,000.
Step 2.	How many select fingerlings should be stocked in the food fish pond, if mortality is 10%?If mortality is 10%, survival is 90%: Number of fingerlings stocked = number harvested/ survival rate = 2,000/0.9 = 2,223.
Step 3.	How many fingerlings must be harvested from the fingerling pond to produce 2,223 select fingerlings, if you save the top 50%? Number of fingerlings harvested = number saved/ percent saved = 2,223/0.5 = 4,446.
Step 4.	How many fry must be stocked in the fingerling pond to produce 4,446 fingerlings, if mortality is 50%? If mortality is 50%, survival is 50%: Number of fry stocked = number harvested/percent survival = 4,446/0.5 = 8,892.

Step 5. How big should the fingerling pond be, if it is stocked at 200,000/ha and if you will stock8,892 fry?

Size of fingerling pond = number of fry that will be stocked/stocking rate =8,892/200,000 = 0.0446 which is rounded to 0.045 ha.

Step 6. How big should the food fish pond be, if it is stocked at 7,000/ha and if you will stock2,223 fingerlings?

Size of food fish pond = number of fingerlings that will be stocked/stocking rate =2,223/7,000 = 0.3175 ha which is rounded to 0.318 ha.

The second, third, etc. generation of selection can proceed as described above. If the select brood fish cannot produce enough fish for both the selective breeding programme and for the production ponds, genetic improvements that are being achieved in the select population can be transferred to the production fish beginning with the second generation of selection, if the farmer uses the culls from the breeding programme as brood fish to produce fry for his production ponds. Even though these brood fish were culled, they came from the selective breeding programme, and their parents were select brood fish.

This will enable a farmer to transfer the genetic gain to the farmed fish, but it will be with a one-generation delay. Additionally, when a farmer replaces one generation's select brood fish with their successors, he can transfer the previous generation's select brood fish to the production brood fish population. This will also enable the farmer to transfer genetic gain to the production population. Some of the ways genetic gain can be transferred to the production ponds are illustrated in Figure.

If selection will occur only when food fish are harvested, a farmer will not be able to transfer the genetic improvement to the production population as quickly, unless the select fish can produce enough offspring for both the breeding programme and for the production ponds. If the select brood fish can produce only enough offspring for the breeding programme, a farmer must use unselected brood fish to produce fry for the production ponds after the first generation of selection. After the second generation of selection, he can either use F_1 select brood fish or fish that were culled when he created F_2 generation brood fish. This means a farmer will realise no genetic improvement in his production fish for one generation, but thereafter, he will be able to transfer it with a one-generation delay.

Asynchronous Spawning: If a farmer cannot spawn his fish synchronously, he should divide the population into age cohorts and select for growth rate independently within each cohort. To initiate this breeding programme, a farmer should spawn the fish using normal management techniques. As before, it is better if egg masses can be collected and incubated. If the eggs are usually incubated by the mother in the ponds, females should be closely monitored, and fry should be collected from each female as soon as they hatch or begin to swim away.

Egg masses or newly hatched fry should be grouped into daily (24-hour) age cohorts. If it is not possible to collect enough families within a 24-hour period, the time interval for each cohort can be stretched to 48 hours. Each cohort should be composed of at least 5 families, and there should be at least five cohorts. These numbers are not carved in stone. A farmer does not have to discard a cohort simply because it is made up of only four families. There are two basic premises behind this work plan: the first is that a cohort should be composed of several families; the second is that at least 50 parents (25 males and 25 females) should produce offspring for the breeding programme. As before, if this species is usually produced by a two-phase production process, selection for increased length will occur when fingerlings are harvested and when food fish are harvested. If production is usually a single phase with no fingerling phase, selection will occur only when fish are harvested for food. Individual selection will be used to select the best fish from each cohort.

This breeding programme will require five to ten 0.04-ha ponds. The number of ponds depends on the size of the project (how many cohorts will be produced) and whether production is a two-phase or a one-phase process. The exact size of the ponds can be determined as was described previously and as was outlined in Table 14. The exact size is not that important, but all ponds should be similarly sized. The reason you want small ponds is because a large number can be squeezed into a small space. Additionally, it is cheaper to build a 0.04-ha pond than a 0.1-ha pond.

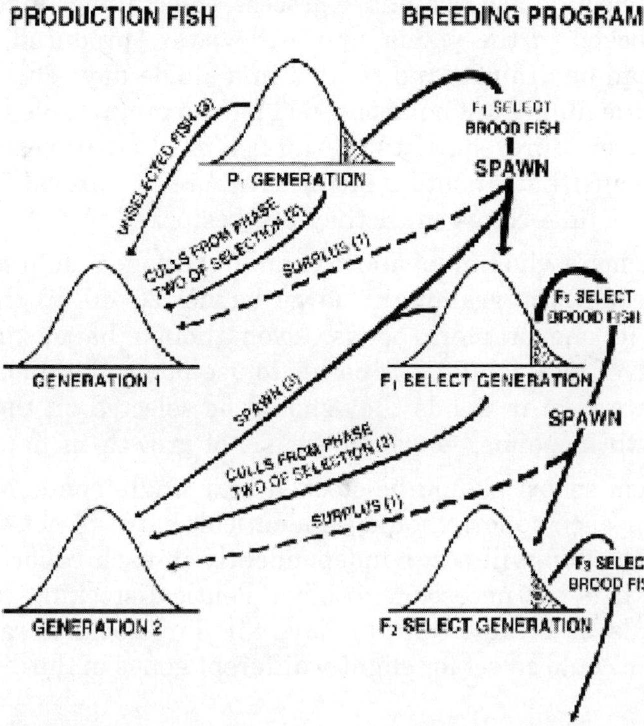

Figure: *Schematic diagram illustrating the ways genetic gain can be transferred from the selective breeding programme to the production fish that a farmer grows for market. If the select brood fish can produce enough for both populations, the transfer will be immediate, and the means of the two populations will be the same (path 1). If selection is a two-phase process, the culls from the second phase can be used as brood fish to produce the production fish (path 2). If this approach is used, some of the genetic gain will be transferred immediately, but the production mean will always lag behind the select mean. If selection is done only at harvest, unselected fish will have to be used to produce the first generation of production fish, and either the culls from the breeding programme or the previous generation's select brood fish can be used thereafter (path 3). This approach will transfer the gains with a one-generation delay if the previous generation's select brood fish are used; if culls from the breeding programme are used to produce the production fish, the production mean will be slightly better than the mean of the previous generation of select fish. All assumptions about mean values were generated using the premise that there is no environmental influence on the phenotype and the trait has a large heritability.*

If selection is a two-phase process, select fingerlings could be restocked in the ponds where they were produced, if the ponds can be drained and refilled in a single day. This would reduce the number of ponds needed for the programme by one-half. The only prerequisites are: all fish must be harvested; the farmer must have holding facilities where he can safely hold the select fingerlings until they are restocked.

Farmers who cannot afford to build the ponds could conduct this breeding programme in large hapas (20–40 m) that are placed in one or more ponds. Even though hapas are less expensive than ponds, it is better to use ponds, because if the fish are grown in ponds they should be selected on the basis of growth in ponds, not on the basis of growth in hapas.

Each cohort should be stocked in a single pond; stocking rates for each cohort should be identical, but it is not critical, since selection will occur independently in each cohort. Even though it is not necessary to have identical stocking rates in all ponds, it is advisable to have similar stocking rates, or selection could select for slightly different genes in the different cohorts.

The best way to create the population of fish that will be stocked in each pond is to choose an equal number of fish from each family within a cohort. The fish chosen from each family must be randomly selected; they cannot be chosen because they are the largest, etc. Therefore, families must be isolated until family size is equalized. If the families within a cohort are mixed before the proper number is chosen to create the desired stocking rate, the largest family will be over-represented and the smallest will be under-represented.

When the fingerlings are harvested, the top 35–50% from each cohort (pond) should be stocked in a single grow-out pond. To determine the fingerling cut-off value for each cohort, a sample of 100–200 fingerlings from each cohort should be measured to the nearest millimetre, and the phenotypic value that corresponds to the desired cut-off percentage should be determined, as was outlined in Figure. The cut-off value must be determined independently for each cohort, because they are managed as temporary, separate sub-populations during the selection process, and selection will occur independently in each cohort.

As before, the intensity of selection is not that critical, but it must be the same for each cohort. Fish from different fingerling ponds (cohorts) should not be mixed when they are harvested and measured. After selection, the select fingerlings from each cohort should be stocked in a separate food fish ponds. Fish from different cohorts should not be mixed.

As was the case when stocking the fingerling ponds, the stocking rates for the food fish ponds should be the same or at least similar. The select fingerlings in each cohort should be grown using normal production techniques. Just prior to harvest, a random sample of 100–200 fish from each cohort should be measured to determine the cut-off value for each cohort. The top 10–20% from each pond (cohort) should be saved to form the population of select brood fish. Again, the intensity of selection is not that critical, but it must be the same for all cohorts.

Once the select brood fish have been chosen from each cohort, they can be mixed and stocked into one or two select brood stock ponds. These ponds should contain no other brood fish. As before, a major goal of the selection process is to have at least 100–200 select brood fish. The two-phase selective breeding programme is outlined in Figure.

The brood fish should be managed and used to produce fry for the breeding programme and for the production ponds, as was described in the previous sub-section.

If this selective breeding programme were going to be conducted by a scientist at a research station, a major aspect of his experimental plan would be to make all aspects of management identical for all cohorts. But a farmer does not have to worry about this. It would be nice if all cohorts were managed identically during each phase of selection, but because each cohort is stocked in a single pond and because selection will occur independently within each cohort, minor management differences among the cohorts will not affect selection. For example, if one food fish pond is stocked with 5,000 fingerlings/ha and the others are stocked with 4,000 fingerlings/ha, the different stocking rates could affect the average growth rate of the cohorts, but since selection will occur independently in each pond (cohort), it probably will not affect the selection process.

However, large differences in management among the ponds could affect the outcome. For example, if one food fish pond is stocked with 15,000 fingerlings/ha and if cattle manure is used as the sole source of nutrients while the others are stocked with 4,000 fingerlings/ha and if fish are fed rice bran, selection may select for different genes in the differently managed cohorts.

Selection for Growth Rate and Another Phenotype

The breeding programmes that have been outlined can be expanded to include another phenotype. If a farmer wants to improve phenotypes other than growth rate, he can easily add traits such as body conformation and/or harvestability. Farmers should use independent culling or modified independent culling to simultaneously improve two phenotypes. Aquaculturists at fingerling production centres should also use independent culling; however, they could use a selection index if they are technically sophisticated and have the expertise and labour needed to conduct this programme.

When independent culling is used, a farmer needs to determine the overall intensity of selection in order to calculate the cut-off percentage for each trait. This process was described in Chapter 4. A farmer should try to save 10–20% of the population at the food fish phase of selection. The phenotypic values that correspond to the desired cut-off percentages should be determined as was described previously and as was outlined in Figure.

Body conformation is an important phenotype, and improving this trait can increase yields. A deeper-bodied fish or a thicker-bodied fish will carry more muscle on its frame, which means each fish weighs more per centimetre body length than a normal, streamlined fish.

Selecting for weight might improve body conformation, but this type of selection does not guarantee that body conformation will be improved. Selecting for weight simply improves average weight, and heavier fish might be fish that are longer, that have larger heads, etc. Improving body conformation has been a continuing goal of many cattle, swine, sheep, and poultry breeding programmes, because animals with better body conformations produce more meat.

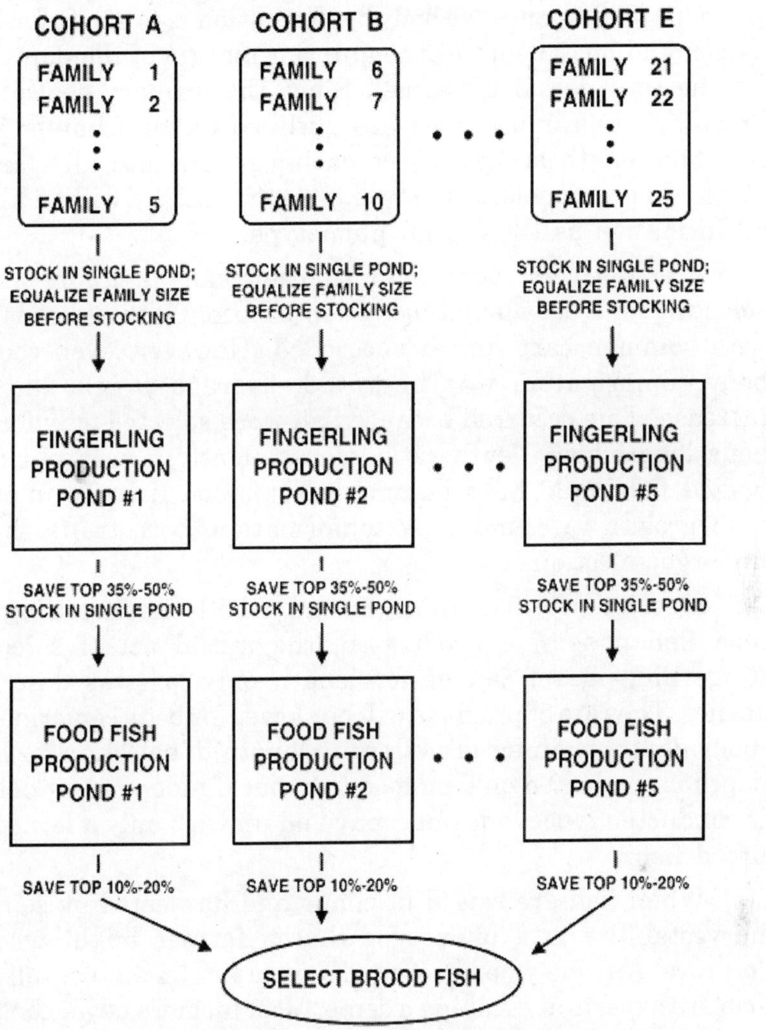

Figure: *Schematic diagram of a selective breeding programme that uses individual selection to improve growth rate when a farmer cannot synchronize spawning. This selective breeding programme divides the population into age cohorts, and selection occurs independently within each cohort. In the programme outlined in this figure, there are five cohorts (A-E), and there are five families within each cohort. If there is no fingerling phase, step one can be eliminated, and the selective breeding programme can be conducted in five ponds. If the number of families can be increased in each cohort, fewer cohorts will be needed, which will decrease the number of ponds that are required. The fate of the culled fish was described in Figure.*

One way to improve body conformation is to select for both length and body depth at the anterior margin of the dorsal fin (at the first dorsal fin spine). Since the primary goal of the breeding programmes that are outlined in this chapter is to select for length as a way of increasing growth rate, all a farmer has to do is add selection for body depth, and he will add body conformation as the second phenotype.

Alternatively, body conformation can be improved by selecting for a body length:body depth ratio. This has been done with common carp and it succeeded. However, even though body conformation was improved, average weight did not increase. This occurred because fish were selected only for the body length:body depth ratio. Consequently, small, but deep bodied fish could have become select brood fish. To improve both growth rate and body conformation, both traits should undergo selection.

If selection will occur at the end of both the fingerling and food fish phases of production, the initial act of selection (fingerling phase) can be for length only, as was described earlier. The rate of progress will be slower for body conformation (body depth) than for growth rate (length) if selection for body depth is conducted only on food fish, but if independent culling is conducted twice, a farmer may end up with only a few select brood fish.

When fish are raised in ponds and harvested by seining, harvestability is a phenotype that a farmer might wish to improve. Anyone who has seined a pond knows all too well that fish are experts at escaping a seine. Most farmers never consider the costs of harvesting when they determine annual production budgets, but it can be significant in terms of labour and equipment. Hard-to-capture fish can become stressed and killed as a result of repeated seinings. Finally, fish that are not captured cannot be sold or eaten.

If a farmer wants to produce fish that are easier to capture, he can add this phenotype to his selective breeding programme and select for both growth rate (length) and harvestability by using independent culling. If a farmer decides to select for harvestability, he needs to define it as "fish that are captured during the first seine haul." If this is done, independent culling will be a two-step process: The initial step will be to save only those fish that are harvested during the first seining; all other fish will be culled. The second step will be to select for length in the fish that were saved (seined).

It may be that so few fish are harvested during the first seining that a farmer cannot select for length efficiently or meaningfully. If this is the case, the goal should be modified so that fish that are captured during the first two seinings are saved.

The ability to escape a seine is not only a detrimental phenotype in that it increases production costs, it is detrimental because it can produce slower-growing fish, if a farmer conducts selection for growth rate improperly. If a farmer cultures fish using multiple batch production, selection for growth rate can be done only during the first harvest after the pond is filled.

Thereafter, size and age are confounded, especially if fingerlings are stocked to replace fish that are harvested or if fish can reproduce in the pond.

If a farmer chooses to select for growth rate and for a second phenotype, growth rate will be improved more slowly than it would be if it were the only phenotype under selection. A farmer could select for growth rate, harvestability, and body depth (or any other phenotype he wants to improve) by using independent culling, but the rates of improvement for all three traits would be small.

Fundamentals of Hybridization in Fish Culture

Hybridization has become a common practice in fish culture, though it has not yet attained its rightful place in this branch of economy. The hybridization of fish for practical purpose has demonstrated its economic value. In the U.S.S.R. hybridization has solved the problems of carp culture in the northern regions. New productive forms have been successfully created by crossing species and varieties of Carassius, as well as by crossing carp and Carassius. Research on crossing of herbivorous fishes for their acclimatization in new areas has been initiated. Economic characteristics of a number of hybrids of ganoids have been determined. Experiments have shown that hybridization has extensive possibilities in sturgeon culture.

One may ask why hybrids should be created when excellent non-hybrids forms, sturgeon for example, are available. Hybrids are offered along with the pure species and not as substitutes.

Preferences for hybrids are based on inherent characteristics of hybrid heterosis (hybrid vigour), which manifest as intensive growth rate, higher viability, adaptive flexibility, and sometimes early sexual maturation.

Charles Darwin attached great significance to hybridization: "If we do not take it for granted, at least we shall consider it highly probable - the existence of a great law in nature, the law of crossing animals and plants which are not closely related to each other, and this is extremely useful and even necessary".

Natural Hybridization

Study of causes and regularities of hybridization processes occurring in nature reveals possibilities and methods of utilisation of hybridization if fish culture. It should be noted that in no other group of the animal world but fish do remote (interspecific and intergeneric) crossings occur so frequently in nature, producing numerous hybrids.

This is particularly true of fresh water fishes, as reproductive isolation is violated more frequently among them than among sea fishes because of less constant hydrological conditions in rivers as compared to the seas.

According to the literature available to us, 212 different fish hybrids have been recorded up to 1956, of which only 30 are of sea fishes. The number of hybrids in nature may be much higher, but not all of them are known to investigators or identified correctly.

In nature, hybrids occur mostly as single specimens, but sometimes large scale hybridization may occur as a result of violation of reproductive conditions of the species. Abundance of individuals of one species occurring along with less numbers of other species may result in an increase of interspecific crossing. That is why the probability of crossing is high when large numbers of a species are introduced into a new area, while it remains low in the aboriginal species.

Fish hybrids are of interest from two points of view:

i. as objects for commercial fish culture when heterosis of the first generation is to be used; sterile hybrids are suitable for this purpose; and

ii. as initial material for producing new hybridogenic forms when the ability for reproduction is a necessary condition. The problem of hybrid reproduction is the most important problem in hybridization.

Great variations in the reproductive ability of hybrids are observed: from complete fertility of intergeneric hybrids of some Cyprinodontidae to complete or incomplete fertility of interspecific and intergeneric hybrids of the sturgeon (Acipenseridae), the carp (Cyprinidae) sunfishes (Centrarchidae) and others.

The degree of fertility of hybrids does not always correspond to the degree of taxonomic relationship of the species crossed. In fact, intergeneric hybrids may be more fertile than intrageneric ones. Sometimes, even reciprocal hybrids produced by crossing the same species, but by substitution of a female of one species by a male of another, and vice versa, may be different in fertility.

Individual hybrids may vary in fertility. In fact, some individuals of so-called sterile hybrids can produce offspring. Therefore, one of the methods of eliminating complete sterility in such hybrids is to carry out hybridization on a larger scale to provide a larger number of hybrid individuals, some of which may be fertile. Fertility may be expected to increase in the second hybrid generation. This is true of such hybrids as Cyprinus x Carassius which, as a rule, are sterile; but among them some exceptional individuals may be fertile. Investigations carried out by A.I.Kuzema has shown possibilities of successful selection work on Ukrainian hybrids of Cyprinus x Carassius.

Fertility of most hybrids is disordered to a certain extent, and they are sometimes sterile. In the process of evolution of various species, they became gradually isolated ecologically, geographically and in other respects. Due to isolation a sort of barrier appeared which prevented them from free crossing. However, such an isolation was not always reliable; the possibility of interspecific crossing still remained. This resulted in the development of a second barrier - disorder in hybrid fertility, preventing them from mixing and destroying specific differences. But where the first barrier acts without failure, fully eliminating the possibility of two species crossing, the second barrier might not appear.

If such species are crossed artificially, the first barrier is removed, and a fertile hybrid is produced. Charles Darwin meant these very cases when he wrote "there exist species which are not easy to cross, but their hybrids, once produced, are highly fertile". Our experiments have shown that man, having succeeded in crossing such species, may find access to reproductive potentials that do not occur in nature.

Chapter 7

Artificial Hybridization

It has been possible to produce fertile hybrids of beluga (Huso huso) and sterlet (Acipenser ruthenus), two species which do not cross in nature due to marked ecological differences.

The sturgeons are remarkable for their ability to cross and for their hybrids that are fertile to various degrees. For many years we crossed the sturgeons in most specific combinations in order to produce interspecific and intergeneric hybrids, not only in the first generation, but also in the second and even the third generation (also back-cross and triple-cross). In our work we did not come across any instance in which it was impossible to produce viable offspring.

Without detailed cytogenetic analyses of the causes of fertility and sterility of fish hybrids, I would like to point out that there are more chances of producing completely fertile hybrids if the species crossed have equal chromosome number, which must be homologous and able to conjugate in the course of gametogenesis. When there is unequal chromosome number and violation of the conjugation process, sterility occurs in various degrees.

Until recently the sturgeons and their hybrids were not studied cytogenetically. We have initiated research work on chromosome complexes of parent species and their hybrids, which has already given interesting results. The various species of the sturgeon family such as Huso huso, sterlet (Acipenser ruthenus), starred sturgeon (A. stellatus), ship sturgeon (A. nudiventris) and, probably, some others are characterized by a small number of chromosomes (about 60), whereas the chromosome number of A. guldenstadti is more than twice as much. Therefore, crossing of the first four species in any combination is likely to produce fertile hybrids; crossing each of the four with A. guldenstadti produces hybrids which are either of limited fertility or are completely sterile, the results of abnormal gametogenesis. Hence, before starting work on hybridization, it is necessary to study chromosome complexes of the species to be crossed; this will permit prediction, with a greater degree of probability as to whether or not the hybrids will be fertile or sterile, and will avoid waste of labour and money.

Heredity of Hybrids

Many scientists, I.V. Mitchurin in particular, have shown that development of some characteristics of hybrids, to a great extent, depends on the environment in which the hybrids were bred. This environmental selection of characters is possible due to the great heterozygosity of the hybrid.

Study of morphological characteristics of certain hybrids of the sturgeons, the carps and others (as compared to their parent species) shows that hybrids have an intermediate heredity not only in the first generation, but also in the second, as well as in the generations produced by back-crossing and triple-crossing. In comparison to the first generation, the following generations, often but not always, are characterized by greater variability; however, no typical morphologic segregation followed by return to the initial species has been observed. Hence it was believed that interspecific hybrids possess the so-called permanent intermediate heredity, which does not obey Mendel's law of segregation.

But recent experimental research has proved that the type of heredity observed in crossing different species does obey the law of segregation. But the law, revealed by the behaviour of genes, is impossible to observe visually in morphological characteristics. The species crossed are different in many characteristics, and consequently in a great number of genes. In interspecific crossing high polymeric inheritance occur when numerous genes affect the development of some definite characteristics in a similar way. This is why in the second generation a much greater number of combinations is inevitable; and among the mass of intermediate specimens, only a few of them are homozygous, which, in all characteristics show a return to the parental form. However, often such individuals are not found in the experiment.

In intraspecific crossing, (inter-racial, for example) it does not matter which race is represented by the male or female, since the characteristics of the offspring are not affected. That is to say, reciprocal hybrids do not differ from each other.

This is expressed by the formula: $A \times B = B \times A$. In interspecific, intergeneric and more remote crossings, reciprocal hybrids may be different, as for example the mule and the hinny which are the reciprocal hybrids between horse and the donkey.

Differences in reciprocal forms of some fish hybrids are revealed in the following: (i) viability, (ii) growth rate, (iii) rates of development of other characteristics of heterosis and (iv) morphological characteristics. Marked difference in viability is observed in reciprocal intergenetic hybrids of crucian carp (Carassius carassius); the offspring of Carassius females crossed with males of other genera of Cyprinidae (for example Carassius @& x Tinca B&) are viable, whereas the reciprocal offspring (of Carassius males) are not viable. Considerable differences in growth rates of fry were found in reciprocal hybrids of the sturgeons. The hybrid possessed some features characteristic of the maternal species, revealing matroclinic inheritance. Morphologic differences between reciprocal hybrids are even more pronouced: in countable characteristics (number of vertebrae, fin rays) these differences are again matrilinear.

Differences of reciprocal forms of hybrids are determined by cytoplasmic differences of the species crossed, resulting in different interaction of the nucleus and the plasma in reciprocal crossing. Genetic predetermination of cytoplasm may lead to incompatibility of the cytoplasm in certain species with chromosomes of some other species. As a consequence, some reciprocal hybrids may be viable, whereas some others may not be. The same is true of the factor determining the manifestation of various degrees of heterosis in reciprocal hybrids. Cytolplasmic heredity is a basis of matroclinic inheritance as well.

The nature of many phenomena in remote hybridization cannot be explained without consideration of gynogenesis as one fundamental problem of the theory. Gynogenesis is a form of sexual reproduction, when in the process of fertilization, the sperm penetrates the ovum and activates it for development without contribution of paternal chromosomes, so that heredity in the offspring is determined by the female pronucleus alone.

As a result, pure matroclinal pseudohybrids are produced, represented as a rule by females only. Some scientists consider gynogenesis to be a special case of parthenogenasis. This is not correct because in parthenogenesis, activation of the ovum is not initiated by sperm but by some other agents.

Gynogenesis occurs in nature and can be demonstrated experimentally. Most interesting instances of natural gynogenesis are Carassius auratus gibelio and the Mexican viviparous fish Mollienesia formosa. They reproduce gynogenetically, spawning with males of some other species.

Some experiments on remote hybridization have resulted in so-called hybrid gynogenesis. Our numerous experiments in crossing Carassius carassius females with Leuciscus cephalus males (different subfamilies) have shown that only a small part of the offspring was viable, and these were matrilinear, i.e. gynogenetic due to the exclusion of the male chromosome complex during development. Most of the embryos died at the very beginning of development owing to incompatibility of maternal and paternal hereditary substance.

Of great theoretical and practical interest, is radiation gynogenesis, produced when eggs are fertilized by sperm exposed to X-rays in dosages sufficient to stop its hereditary function, but not its ability to penetrate the egg and activate its development. In fact, in the fertilization of the sterlet (Acipenser ruthenus) eggs by the sperm of the sturgeon exposed to X-rays, or the beluga (Huso huso) eggs by the irradiated sperm of the sterlet, completely matrilinear offspring were produced (no paternal characteristics were observed).

Since gynogenesis produces only females it may be possible to utilise it for control of sexual ratio in fishes, which is of great practical importance. Increase in the number of females of the sturgeon, for example, would be desirable both for commercial fishery and for artificial culture. Experiments in this field should be continued on a larger scale.

Heterosis

Heterosis in fish hybrids is a natural phenomenon observed in various families such as Acipenseridae, Salmonidae, Cyprinidae, Percidae, Centrarchidae and Poecilidae. As a rule heterosis is more prominent in intergeneric hybrids which are not closely related, than in interspecific or intraspecific hybrids.

Heterosis is revealed in various degrees. For example, growth rate of the hybrid may exceed that of the parent species (typical heterosis), or sometimes growth rate of the hybrid may exceed that of one of the parent species and be equal to or less than the other parent (incomplete heterosis).

In the second hybrid generation, heterosis usually declines, but in back-crosses of the first generation hybrid with the parent species (in which the characteristics that are of interest to us is more pronounced) the hybrid may exceed both the parent species; that is, the hybrid will show a prominent heterosis. For example, the hybrid of Huso huso x (Huso huso x Acipenser ruthenus) exceeds Huso huso in growth rate.

Both in evolution and selection, hybrid forms are of more value if their fertility is not violated.

Heterosis should be used on a greater scale in selectional culture of new, valuable fish and in acclimatization as well as in crossing of hybrids for commercial culture in ponds and other inland reservoirs.

Control of Fish Diseases by Selection

Selection of fish for resistance to diseases may prove, and in some cases has already proved, to be an important means of control of fish diseases. Selection work often necessitates transportation of fish from one area to another and this often leads to introduction of diseases and parasites into pond establishments and commercial waters where previously they were absent. Selective breeding of animals and plants for higher resistance to diseases assumes ever greater importance. Often resistance to a specific disease is dependent on a relatively simple mutation involving one or a few genes. Any gene mutation is accompanied by change in the synthesis of proteins and may cause protein incompatibility of parasite and host (Kirpichnikov et al., 1967).

Selective breeding for higher resistance to disease has been applied in fish culture only in recent years and the data available are therefore limited. Insufficient study of fish diseases, especially the infectious ones, and contradictory observations on immunological reactions in fish have also contributed to this. Until recently some authors considered that fish were not capable of producing antibodies. The studies of recent years show that fish do produce antibodies but not so intensively as the higher vertebrates. Immunobiological reactions of fish are undoubtedly affected to a great extent by water temperature; the higher the water temperature, the more pronounced the reaction. At very low water temperature the processes of antibody formation are almost imperceptible (Krantz et al., 1964; Vladimirov, 1966).

Bactericidal activity of organs and tissues has been discovered in fish as well. It is probably dependent on the presence of such antimicrobial proteins as complement, lysozyme and properdian (Lukyanenko and Mieserova, 1962; Lukyanenko 1965; Bauer et al., 1965). At temperatures lower than 10°C the mechanism of excretion and bactercidal action of organs and tissues predominate over other immune mechanisms (Avetikyan, 1959). Phagocytosis, a mechanism of antibody formation begins to function with increase in temperature (Goncharov, 1963, 1967). Recent data on fish immunity provide selection breeders with a theoretical background for the breeding of disease-resistant fish strains.

Selective Breeding of Disease-Resistant Fish

When a disease occurs in a stock, it does not affect all fish to the same extent; some fish, sometimes large numbers, are not affected. Schäperclaus (1953) cites this example: In the spring of 1950 a small pond in Peitz fish farm was stocked with 665 two-year-old carp which were kept there until early June. In June, an outbreak of infectious dropsy occurred. When the pond was fished out, it was found that 4 percent of the fish were either dead or doomed to die; 54 percent of the fish had external ulcers typical of the disease; and 42 percent showed no symptoms of the disease. Even progeny of a single pair of spawners reared under the same environmental conditions did not all possess the same resistance. Individual specimens may serve as starting points for selection work.

Fish belonging to different strains and forms of any species differ in their resistance to disease. The wild carp (Cyprinus carpio) and especially its Amur sub-species (C. carpio haematopterus) are known for their higher resistance to infectious dropsy than the cultured carp (Karpenko and Sventycki, 1961). Amur wild carp × cultured carp hybrids also possess high resistance. The replacement of cultured carp with hybrid forms in recent years in a number of Ukrainian fish farms have resulted in considerable decrease in the incidence of dropsy in this region.

The same was observed in relation to another widespread infectious disease of carp, the air-bladder disease; it is widespread, not only in the USSR but in the other European countries. Arshaniza (1966) carried out experiments on joint rearing of the cultured carp and the hybrid of the fourth generation of cultured carp × Amur wild carp. The results showed that the incidence of airbladder disease amounted to 78 percent in the former, while it occurred in only about 30 percent of the latter. Symptons of illness were more pronounced in the cultured carp than in the hybrid.

Considering these examples, we may assume that work on development of disease resistant strains of fish may be started not only with the selection of specimens possessing higher individual resistance (fish that remained healthy within a stock affected by the disease), but also with the development of breed groups of hybrid origin, distinguishable by higher resistance to a particular disease. Naturally, such selection work should be carried out at those fish farms which are always under the effect of the disease. Otherwise the increased disease resistance of the selected stock would easily be lost.

As stated earlier, data are scant on the selection of disease resistant stocks of fish, and many of the studies undertaken on the problem have either not been carried through or are still in progress. Only three examples of such studies have been carried to conclusion, two of which concern infectious dropsy. For a long time dropsy was considered a bacterial disease, Aeromonas punctata being its pathogenic agent (Schäperclaus, 1930, 1954). Subsequently, many scientists have questioned the bacterial nature of the disease and have come to the conclusion that dropsy is caused by virus.

Direct evidence of the viral nature of dropsy was obtained several years ago by Yugoslavian scientists (Tomasec et al., 1964). They succeeded in growing the causative agent of dropsy on artificially-cultured tissue of carp kidney, and infecting healthy fish with it. The second International Symposium on Fish Diseases held in Munich in 1965, recognised a virus as the primary pathogenic agent of dropsy, while A. punctata and other saprophytic bacteria complicate the course of the disease.

Breeding of Dropsy-Resistant Strain of Carp

Attempts to create a stock of dropsy-resistant carp were made for the first time by Schäperclaus (1953). Work on selection of dropsy-resistant carp was started by him in 1935. Carp not affected by the disease, in ponds where dropsy outbreak had occurred, were chosen for selective breeding. For two successive years the carp were exposed to large amounts of A. punctata with a view to artificially strengthening their natural immunity.

Fishes having even the slightest symptoms of disease were eliminated. The offspring of both selected and non-selected fish were compared for dropsy resistance; the results showed that of the former only 2 percent contracted dropsy, but of the latter 30 percent did. No dropsy was recorded among the progeny of the selected spawners. From 1939 to 1943 the mortality rate for the ponds (65 ponds with a total area of 180 hectares) where the progeny from selected spawners was reared, averaged 11.5 percent, its range varying between 0 and 37 percent. The average mortality rate for the control ponds (76 ponds with a total area of 474 hectares) was 57 percent, its range being 4 to 96 percent. Decrease in mortality was observed in subsequent years as well. From these observations Schäperclaus concluded that resistance to dropsy is inheritable and it is possible by means of correct selection to obtain stocks possessing relative immunity to this disease.

However, it is possible to obtain this effect only where a selected stock is not mixed with any other and where no infection is brought in from outside, either with the water or with the fish delivered to the station. Experience over many years has shown that infection brought into a specially selected stock manifests, as a rule, in an acute form.

But such complete isolation of the selected stock cannot be maintained in the carp farms situated within the region of the natural habitat of Cyprinus carpio. Infection is continually brought in either by the wild fish penetrating into the fish farm, or by the water supply or perhaps by some other unknown agent. It was suggested by Kirpichnikov et al., (1967) that such fish farms should select spawners from breed groups that are most resistant to dropsy and increase their resistance further by means of controlled selection. Such work has been started by Kirpichnikov at the fish farms of the Northern Caucasus. The aim of the first phase of this work was to compare the resistance to dropsy of different breed groups of carp. The following groups of carp have been tested, singly as well as in mixed groups.

i. Ropsha carp (hybrid Amur wild carp × mirror carp, F_4 and F_5),
ii. Ukrainian carp selected by A. Kuzema,
iii. local scaly carp and low-bred mirror carp of unknown origin,
iv. wild carp of the Don river.

The trials for resistance to dropsy were carried out at one of the fish farms. The larvae were reared separately. Fish of different breed groups were combined and reared for three years (1963–1965). The fish were marked by fin clipping; however, under the conditions of a hot summer, the fins grew, causing errors in differentiation of breed groups. The results of the test showed that the various groups showed marked differences in their resistance to dropsy. The Ropsha carp, local scaly carp and Don River wild carp proved to be the most resistant to this disease. The enhanced resistance of the above groups is apparently connected to their wild carp origin. Besides high resistance to dropsy, the Ropsha carp was characterized by generally higher viability, and this carp was therefore selected for breeding at the fish farms of the Northern Caucasus, where outbreaks of dropsy often occur. In the spring of 1966, three-year-olds of all groups had attained sexual maturity and served as brood stock of the second generation, which also was subjected to intensive selection for resistance to dropsy. Though this work is far from completion it promises great economic possibilities, especially in the Northern Caucasus, where dropsy is a major hazard in fish culture.

Breeding of Disease-Resistant Trout

In the early fifties, work was started on the creation of trout stocks (the brown trout, Salmo trutta m. fario and the brook trout, Salvelinus fontinalis) resistant to two widespread bacterial diseases, furunculosis and ulcer disease (Wolf, 1954). The initial phase of the work was to test the susceptibility of different fish populations to these diseases. For this, ponds at one trout farm were stocked with fish brought from different bodies of water. Fish affected by the diseases were released into these ponds and the dead fish were fed to the experimental ones. The diseases broke out in all experimental ponds, causing mortality. In the first year, the mortality rate of the brown trout in the experimental ponds varied from 35.7 to 65.8 percent; that of the brook trout fluctuating between 5.3 and 99.6 percent. Thus the two species were seen to differ in their resistance to the diseases. Populations with high resistance were chosen for further selection work. Results obtained in subsequent experiments are not available in the literature accessible to us.

Further work on selection of trout for resistance to furunculosis has been reported by Elinger (1964). Diverse geographical strains of the brown trout and the brook trout were intensively exposed to Aeromonas salmonicida, and the less resistant individuals were removed by mass selection. Survivors were selected as brood fish for the next generation, which was similarly exposed to the disease. Hybridization through crossing more promising strains in some instances improved resistance. Resistant strains were compared with regular hatchery stock under actual hatchery conditions. Results obtained thus far indicate that resistance to furunculosis is being achieved. The work is in progress.

Effect of Selection Work on the Distribution of Parasites and Diseases of Fishes

For carrying out selective breeding of fish it is necessary to obtain fish belonging to different populations, forms, subspecies, or species from different regions for rearing in experimental fish farms. It is apparent that pathogenic agents of different diseases might be brought in with these fish. The danger of bringing in infection is negligible when eggs are chosen as the initial material for selection. There is only one instance on record where infection was introduced into a hatchery with eggs; the pathogenic agent of whirling disease, one of the most troublesome trout diseases, Myxosoma cerebralis, was introduced with trout eggs.

It was supposed that the spores of the organism could stick to the trout eggs and be introduced in this way into new fish farms and bodies of water. However, there is no corroboration of the possibility of such introduction. Efforts in obtaining carp eggs under hatchery conditions, with a view to propagate them artificially, have shown that healthy progeny may be obtained from the eggs of diseased fish. It has been demonstrated that such widespread disease of carp as infectious dropsy and air-bladder disease are not transmitted germinatively. However, there exists the possibility of introduction of parasites infesting the eggs into new bodies of water and fish farms. Fortunately, such parasites are few. Only one species of Coelenterata, infesting acipenserid eggs (Polypodium hydriforme) has been recorded in the U.S.S.R. The danger of bringing it along with eggs is not excluded, and every precaution should be taken while transporting eggs for fish-cultural purposes.

The danger of possible introduction of pathogens while transferring fry or older fish is more real. Soviet ichthyopathologists have collected a good deal of data on this. The following is one example. From 1958 to 1962, large numbers of fry of the grass carp Ctenopharyngodon idella, caught in the rivers of China, were transferred to the U.S.S.R. In spite of all the precautions taken, more than 20 species of parasites were introduced with these fish. Many of these parasites cause severe epidemics, not only among phytophagous fish, but among other cultured and commercial fish as well. Thus the cestode

Bothriocephalus gowkongensis came to be introduced into the fish-farms of the European part of the U.S.S.R. and Central Asia; then it infected the carp and many other fishes, causing high mortality. It has since penetrated into many bodies of water (Musselius, 1967), and has been introduced into Ceylon (Fernando and Furtado, 1962) and into Romania and other countries of eastern Europe (Radulescu and Georgescu, 1964).

There are many examples of accidental introduction of pathogenic agents of fish diseases along with fish brought in for selection work. The selection of the cold-resistant strain of carp is based on the crossing of Amur wild carp and the cultured carp (Cyprinus carpio × C. carpio hematopterus). In the course of this work a large number of parasites of Far-Eastern origin came to be introduced into the various fish farms.

Back in 1936 the first lot of Amur wild carp was brought in, resulting in the introduction of two monogenetic trematodes which parasitize the gills, and a cestode. Dactylogyrus extensus and Khawia sinensis cause severe diseases, and at present it is difficult to find a fish farm where these do not occur.

In 1949 some Amur wild carp were introduced into fish farms in the northwestern regions of the U.S.S.R. and Byelorussia. Some of these fish were infected with the parasitic protozoan, Ichthyophthirius multifiliis, which was unknown in the carp ponds of the U.S.S.R. Long-distance transportation which lasted about 20 days promoted the development of further infection and when transferred to fish farms, the wild carp passed on the infection to the local spawners. In the spring of 1950 a heavy infection caused high mortality among the spawners. At the Velikolusky fish farm 100 percent of the spawners were lost, and the mortality in the Byelorussian farms was not less than 50 percent. The disease was introduced with the stocking material into the other farms of Byelorussian and the northwestern regions of the U.S.S.R., and by the mid-fifties it had spread to the fish farms of the Ukraine, Central Asia and Kazakhstan.

It was proved experimentally that fish exposed to infection for the first time turned out to be carriers of the largest numbers of Ichthyophthirius. The intensity of disease is 20 times less in fish exposed to the infection repeatedly and further exposures helped to develop immunity to the disease (Bauer 1955, 1958). Toward the end of the fifties a new lot of the Amur wild carp was introduced into three fish farms of Latvia for selection work. In a year a severe outbreak of dropsy occurred. It was also found that a large nematode, previously unknown in the U.S.S.R., was brought in with that lot of fish. Vismanis (1962 and 1967) described the nematode as a new species, Philometra lusiana. Large females of this nematode, parasitizing under carp scales, cause the formation of ulcers and consequent deterioration of the market quality of the fish.

In 1966 a new lot of Amur wild carp was transferred to Ropsha, an experimental fish farm of the State Research Institute of Lake and River Fisheries; every precaution had been taken. In the autumn of 1967, myxosporidians of the genus Sphaerospora were found on the gills of carp in a number of ponds; this parasite had never been found in Ropsha ponds before.

It is on record that fish in a number of fish farms were infected by air-bladder disease, introduced there with the carp broodstock brought in for selection purposes.

Thus any transfer of fish for the purpose of selection is always accompanied by the danger of introduction of pathogens, even though every precaution is taken.

Only introduction of fish at the egg or at the larval stage (during the first few days after hatching) eliminates this danger. This is possible with adequately worked-out methods of artificial spawning. For salmonids and whitefish such methods have been developed since the close of the last century and, as a rule, these species are transplanted at the egg stage. Only this may explain the fact that not a single pathogen has been introduced with rainbow trout exported from America to Europe and other continents. At present there are well-developed methods for obtaining fertilized eggs of Acipenseridae and phytophagous Cyprinidae. The eggs of the latter are pelagic. Until recently no effective method was known for taking and incubating adhesive eggs as those of carp. Removal of the adhesive material of carp eggs by any of a number of recently developed methods would enable artificial reproduction of this species, as well as long-distance transportation of the eggs and larvae. In recent years the possibility of obtaining healthy progeny from carp spawners affected by infectious diseases and parasites has been examined. Thus in 1965 (in Byelorussia) and 1967 (in Yazhelbyzy) eggs were obtained from carp spawners affected by air-bladder disease. These eggs were incubated in sterile water, and the larvae were released into ponds where the disease had not been recorded, or to carefully disinfected ponds supplied with water from a non-infected body of water. When mass examination of fingerlings was carried out, not a single individual was found affected by the disease.

Artificial reproduction of wild carp has been carried out at the Tsymlansk hatchery-nursery for a number of years and in 1967 about 70 million larvae were obtained. Observations have shown that if larvae are released into carefully disinfected ponds, they will be practically free of parasites when fished out in autumn. Transporting of fish at the egg and larval stage delays the process of obtaining progeny from the transferred fish by several years. But the delay is worthwhile for it excludes the danger of introduction of infectious diseases with fish brought in for selection purposes.

Rine Biotechnology : Bioactive Natural Products and Their Applications

Marine (or blue) biotechnology encompasses the applications of biotechnology tools on marine resources. Marine biotechnology encompasses those efforts that involve the marine resources of the world, either as the source or target of biotechnology applications. Biotechnology is the application of science and technology to living organisms, as well as parts, products and models thereof, to alter living or non-living materials for the production of knowledge, goods and services'. In the case of marine biotechnology, the living organisms derive from marine sources.

Biotechnology is defined as the industrial use of living organisms or biological techniques developed through basic research; marine biotechnology is an emerging discipline based on the use of marine natural resources. The oceans encompass about 71% of the surface of our planet, but over 99% of the biosphere (since organisms are found throughout the water column), and they represent the greatest extremes of temperature, light, and pressure encountered by life. Adaptation to these harsh environments has led to a rich marine bio- and genetic-diversity with potential biotechnological applications related to drug discovery, environmental remediation, increasing seafood supply and safety, and developing new resources and industrial processes.

Drug discovery represents one of the most promising and highly visible outcomes of marine biotechnology research. Biochemicals produced by marine invertebrates, algae and bacteria, are very different than those from related terrestrial organisms and thus offer great potential as new classes of medicines. To date, examples of marine-derived drugs include an antibiotic from a fungi, two closely related compounds from a sponge that treat cancer and the herpes virus, and a neurotoxin from a snail that has painkiller properties making it 10,000 times more potent than morphine without the side effects. However, there are several more marine-derived compounds currently in clinical trials and it is likely that many more will advance to the clinic as more scientists look to the sea for these biotechnological uses.

In addition to new medicines, other uses for marine-derived compounds include: cosmetics (algae, crustacean and sea fan compounds), nutritional supplements (algae and fish compounds), artificial bone (corals), and industrial applications (fluorescent compounds from jellyfish, novel glues from mussels, and heat resistant enzymes from deep-sea bacteria).

Marine biotechnology may include techniques such as bioprocessing, bioharvesting, bioprospecting, bioremediation, using bioreactors etc (so called process biotechnology techniques); aquaculture/fisheries; gene, protein, or other molecule based techniques; while applications may include: health, food, cosmetics, aquaculture & agriculture, fisheries, manufacturing, environmental remediation, biofilms and corrosion, biomaterials, research tools etc. Therefore, marine biotechnology has a horizontal scope encompassing very different applications, for all of which the marine environment is providing the resources. This means everything from deriving a new cancer treatment from a deep-sea sponge to developing an innovative buoy system for monitoring ocean pollution. Like the broader field of biotechnology, marine biotechnology (marine biotech or MBT) can take very traditional forms, such as localized seaweed farming, and high-tech forms. The full range of genomic tools, for instance, is now being applied to such goals as determining precisely how a promising compound derived from a marine organism kills cancer cells. Commercial applications of marine biotechnology includes bio-prospecting, improving the production of marine organisms, production of novel products, particularly food and feed products and diagnostics and biosensors. Bioprospecting includes biotechnology use in whole drug/molecule development process, i.e. screening, identification, efficacy testing, safety testing, large scale commercial production. Bioactive natural products are secondary metabolites produced by organisms that live in the sea. These products have received increasing attention from chemists and pharmacologists during the last two decades. These products have been exploited for variety of purposes including use as food, fragrances, pigments, insecticides and medicines. Through improved biological screening method the role of these products in drug discovery has been greatly enhanced in the last few years. These products show an interesting array of diverse and novel chemical structures with potent biological activities.

Research into pharmacological properties of marine natural products has led to the discovery of many compounds considered worthy of clinical applications. There are great potential in bioprospecting from the sea and marine natural products research has just started to bloom. Today, marine sources have the highest probability of yielding natural products with unprecedented carbon skeletons and interesting biological activities. Many more prospects regarding new habitats for example deep ocean samples and symbiotic systems are still wide open for research.

Bioprospecting

It is high-throughput screening for novel compounds, especially drugs (other uses includes in foodstuffs, nutraceuticals, adhesives, paints, cosmetics, environmental remediation, research etc). It includes biotechnology use in whole drug/ molecule development process, i.e. screening, identification, efficacy testing, safety testing, large scale commercial production. Traditionally only 1 out of 10,000-20,000 molecules extracted from terrestrial micro-organisms, plants or animals finally reached the market, which may take 10-15 years and cost up to $800m12,13. This has resulted in large pharmaceutical groups abandoning their search for new drugs derived from natural substances. Evidence of complexity in this process is the relatively empty pipeline of pharmaceutical companies. However, marine biota present a better opportunity for encountering successful candidates in view of the large biodiversity, lack of current knowledge and extreme environments. Ara-C and Ara-A (estimated annual worth $50-100 million) drugs were developed from sponges in the early 50s and proved to be commercial success stories. Anti-cancer agents from marine organisms have an estimated value of $1 billion a year. Another good example includes Vent-DNA polymerase which is used as a basic constituent in Polymerase Chain Reaction (PCR). Bioprospecting is applied to all animal and plant phyla living in shallow as well as in deep seabed ecosystems. The advantages of the first include less technical complexity and better economic viability whereas for the latter a larger and perhaps more interesting natural resource base. In general, it is thought that the pace of discovery of new species and products that are potentially useful to pharmacology is higher for marine and microbial than for terrestrial organisms.

Marine Bioprospecting: Mining the Untapped Potential of Living Marine Resources

Compared to the high degree of representation of terrestrial-derived bioproducts, the number of marine natural products that have found their way into hospitals, clinics and pharmacies is thus far small. This has more to do with the relative infancy of the field (compared to terrestrial bioprospecting) than any lack of potential for discovery. In fact, the natural products isolated from marine sources tend to be more highly bioactive than terrestrial counterparts. This is in part because they have to retain their potency despite dilution in surrounding seawater to be effective in the "chemical warfare" that allows organisms such as sponges to ward off would-be predators and animals that might attempt to grow over and smother them. Despite lesser attention paid to marine natural products historically, there are notable marine-derived bio products that are commercially available, including:

- The family of antiviral drug, including Aids treatment AZT, based on a group of compounds (arabinosides) extracted from the sponge *Tethya crypta* more than 40 years ago. This success story from marine biomedicine's first wave represents an annual market of more than $50 million.
- The anti-inflamatory and analgesic pseudopterosins isolated from a Bahamian soft coral (*Pseudoterigorgia elisabethae*). This led to the development of bioproducts now used in Estee Lauder skin care and cosmetics lines and currently worth $3-4 million a year.
- Ziconitide, known by its trade name Prialtýÿ is a synthetic form of a compound extracted from the venom of predatory tropical cone snails (Conus spp.). Prialtýÿ was approved by the FDA in late 2004 as a treatment for severe cases of chronic pain in patients with conditions such as cancer and AIDS. Current clinical results suggest that Prialt is a powerful, non-addictive alternative to drugs such as morphine

Easily overlooked in the success stories is the technical difficulty of collecting and processing diverse marine organisms, screening them for bioactivity, isolating and identifying natural products, securing a sustainable source of the product, and various other hurdles on the path toward developing a potentially profitable bioproduct.

There is a substantial cost associated with each stage of the endeavour. Furthermore, even when no expense is spared, and even if the best and brightest researchers are on the job, there is never a guarantee that the effort will pan out, but the vast potential payoffs, not only financially through potential commercial sales but also at the human level in lives potentially saved or improved, keeps the field moving forward. Marine bioprospecting is often used to describe the process of collecting marine biota for natural product screening. Of course, biomedical MBT investigations are usually carried out by large research institutions and consortiums and not private individuals, so when a drug lead doesn't 'pan out,' personal financial ruin is not typically the result. Still, the fact that only a handful of the 10,000 or so novel chemical compounds thus far isolated from marine organisms have made their way into the drug discovery pipeline says something about the chances of any one compound ever making its way to the drug market.

Moreover, of those compounds that do enter the clinical development pipeline, a large number eventually are discarded to the biomedical boneyard.

As an indication of the investment involved in marine bioprospecting, consider that the UN Atlas of the Oceans reports Japan on its own spends nearly $1 billion on such efforts. Of this sizeable figure, 80% is funded through the private sector.

Specimen Acquisition

A 'typical' marine biotechnology natural products research group will have research objectives centred on any and/or all of the following:

- Collection of marine organisms with promising or as yet unstudied bioactivity
- Detection of bioactivity
- Isolation and identification of bioactive compounds
- Evaluation of the pharmaceutical or other potential of novel compounds

The collection of marine organisms allows assessment of their biomedical or biotechnology potential. The collecting (prospecting) phase of the research is, by definition, a field-based endeavour. The follow-on biotechnology assessment phase is then carried out back in the laboratory in the days, weeks, months, and even years following initial collection.

Collection techniques include:

- Traditional snorkel and SCUBA
- Ship-based collection via bottles, nets, trawls, and benthic grabs, sleds, and dredges
- Collection via ROVs (Remotely Opertated Vehicles) and AUVs (Autonomous Underwhater Vehicles)
- Submersible-based collection

CadalminTM Green Mussel Extract (GMe)

CMFRI was developed new product CadalminTM Green Mussel Extract (GMe) to combat arthritis. CadalminTM. GMe contains 100% natural marine bioactive anti-inflammatory ingredients extracted from green mussel Perna viridis. The product is effective to combat chronic joint pain, arthritis/ inflammatory diseases, and improves cardiovascular functioning. It is an effective green alternative to synthetic non steroidal anti-inflammatory drugs (viz., aspirin containing drugs having undesirable side effects).

The active principle in CadalminTM GMe effectively inhibits inflammatory cyclooxygenase-II and lipoxygenase-V, and its activity was found to be comparable to the drugs available in the market. Consuming CadalminTM GMe will avoid unfortunate side effect of these synthetic non steroidal anti-inflammatory drugs. This product is a blend of nutraceutical and nutritional elements. CadalminTM GMe is designed to find a unique way to prevent the degradation by air, moisture, heat and light and to maximize the activity. The product is free from deleterious trans fatty acids, free radicals/free radical adducts, and low molecular weight carbonyl compounds.

Chapter 8

Vertebrate-Derived Compounds

Squalamine

Squalamine is an aminosterol isolated from the stomach and liver of the spiny dogfish, *Squalus acanthus*, a common New England coastal shark species. (When it was discovered in 1993, the compound was reported to exhibit broad-spectrum antibiotic activity. Squalamine was licensed to Magainin Pharmaceuticals (now Genaera Corporation) for development. Of considerable interest is published evidence suggesting that squalamine exhibits anti-angiogenic activity under certain conditions (angiogenesis is the formation and differentiation of blood vessels).

Neovastat" (AE-941)

Neovastat (AE-941) is a derivative of shark cartilage extract. Rather than being a specific monomolecular compound, AE-941 is a defined standardized liquid extract comprising the < 500 kDa (kilodaltons, a unit of mass) fraction from shark cartilage. It has the anti-angiogenic and antitumor activity. The drug's anti-angiogenic bioactivity suggests it could be a valuable agent for use in patients suffering from multiple myeloma and other hematologic (blood) diseases.

Sponge-Derived Compounds

Bengamides and Derivatives

Two novel seven-membered ring heterocycles, bengamide A and bengamide B, were reported in 1986 isolated from an as yet undescribed Fijian sponge belonging to family Jaspidae.

Since this time, a number of additional compounds from the bengamide class have been isolated, most notably from the Fijian sponge *Jaspis*cf. *coriacea*. Bengamides A and B were initially reported to exhibit *in vitro* toxicity to larynx epithelial carcinoma cells, and to have antibiotic and anti-helminthic activity (against the nematode *Nippostrongulus braziliensis*).

Contignasterol (IZP-94005, IPL576,092)

Contignasterol (IZP-94005) from the sponge *Petrosia contignata* was first reported in the early 1990s. The natural compound itself was deemed a good anti-asthma drug candidate based on performance in vivo animal models. One contignasterol derivative, named IPL576,092, shows promise as an oral asthma medication. More recently, IPL576,092 has entered clinical trials as a treatment for diseases causing inflammation of the eyes and skin.

Debromohymenialdisine (DBH)

Debromohymenialdisine (DBH) is an alkaloid originally isolated from the shallow-water Palauan marine sponge *Stylotella aurantium*. The molecule is intriguing not just for its drug-like properties, but also because it's simple molecular structure has yielded to easy total synthesis in the laboratory. Research suggest the compound acts as a highly selective inhibitor of a specific target cell DNA damage checkpoint enzyme during the G2 phase of the cell cycle. The compound is a promising potential Anti-Alzheimer agent.

Girolline (Girodazole)

Girolline was reported to inhibit protein synthesis in eukaryotic target cells. The compound was of interest because it inhibited protein synthesis at the termination steps of the process rather than at the initiation or chain elongation steps. The simple structure of the compound facilitated the production of a modified synthetic analog.

Halichondrins

A compound given the name halichondrin B, belonging to a chemical family known as the macrolides, was isolated by Uemura *et al.* (1985) from the Japanese sponge *Halichondria okadai*. Initial investigations into the bioactivity of the compound revealed that halichondrin B apparently bound tubilin at a site close to the so-called vinca site and altered tubulin depolymerization.

Hemiasterlins (H-286)

This class of novel marine oligopeptides have been shown to be potent antitumor agents. Representatives of this class of natural products have been isolated from extracts prepared from sponges residing in two distinct genera (*Auletta*; *Siphonochalina*). Three different hemiasterlins with drug development potential (hemiasterlin, hemiasterlin A, hemiasterlin C) have been the subject of chemical and biological investigations. These molecules exhibit cytotoxic and antitubulin activity similar to that seen in the dolastatins. Mitotic inhibition occurs through binding to tubulin at the vinca/peptide region in a manner similar to dolastatin.

KRN7000

It was first isolated from the sponge *Agelas mauritianus*. These compounds, dubbed the agelasphins, were shown to exhibit antitumor and possible immunostimulatory activity.

Lasonolides

The lasonolides are a series of marine natural products under investigation for the treatment of cancer. The compounds were isolated in 1994 by scientists from the Harbor Branch Division of Biomedical Marine Research from the sponge *Forcepia* sp. found in Gulf of Mexico deep-sea habitats. These compounds are very potent and show especially promising properties for the treatment of pancreatic cancer. They kill cancer cells in a different way than most other cancer drugs. The exact mode of action is not yet fully understood, and is an area of active research. In addition to anti-proliferative and antitumor properties, these novel macrolide compounds display antifungal activity as well.

Manoalide

Manoalide was isolated from the sponge *Luffariella variabilis* collected in the Indo-Pacific. It is a member of a chemical family known as the sesquiterpenes. (Although this natural product was originally reported as an antibiotic, follow-up work revealed manoalide possesses promising analgesic and anti-inflammatory properties. The compound works by inhibiting Phospholipase A2 (PLA2), which plays an important role in the inflammation process.

Topsentins

The Topsentins are a class of natural products that have been extracted from several sponge species subsequent to their initial isolation from the sponge *Spongosorites ruetzleri*. These compounds have been shown to have significant anti-inflammatory properties. Although the precise mode of action is not known, the compound has been reported capable of suppressing immunogenic as well as neurogenic (originating in nerve tissue) inflammation. The topsentins may hold promise as an arthritis medication or as additives in anti-inflammatory creams for the treatment of skin irritations.

Dictyostatin

Dictyostatin was first isolated in 1994 by Bob Pettit and his colleagues at Arizona State University from a sample of an unidntified dictyoceratid sponge of genus *Spongia* collected in Jamaica. The compound was later isolated from a lithistid sponge of the family Corallistidae and has been more fully repurified and characterized by researchers at Harbor Branch Oceanographic Institution and elsewhere. Dictyostatin inhibits the growth of human cancer cells and has been shown active against certain Taxol-resistant tumors. Its mechanism of action appears to be prevention of the breakdown of tubulin during mitosis in a fashion similar to the successful cancer drug Taxol.

Laulimalide

Sometimes referred to as the fijianolides, the natural products laulimide and (the significantly less bioactive) isolaulimide were first extracted from the Pacific sponge *Cacospongia mycofijiensis*. The bioactivity displayed by these compounds is as microtubule stabilizing agents potentially arresting the development of target cells.

Cnidarian Derived Compounds

Pseudopterosins The pseudopterosins were isolated from a Caribbean soft coral species called a sea whip (Pseudopterogorgia elisabethae). They belong to a class of compounds known as tricyclic diterpene glycosides. Pseudopterosins have been shown to possess potent anti-inflammatory and analgesic (pain relief) properties. They appear to work by inhibiting the synthesis of eicosanoids, (locally functioning hormone-like substances) in specific white blood cells called polymorphonuclear leukocytes.

Eleutherobin

Eleutherobin was first found in extracts made from the octocoral *Eleutherobia* sp. collected in Australia. The reported bioactivity of this natural product, currently under preclinical investigation, is as a microtubule binding agent similar to the anti-cancer drug taxol.

Sarcodictyins

Sarcodictyins were isolated from two Mediterranean coral species, *Sarcodictyon roseum* and *Eleutherobia aurea*. It acts as tubulin interactive agents.

Helminth Derived Compounds

Anabaseine (Hoplonemertine Toxin)

Anabaseine is a nicotinoid alkaloid. It is capable of stimulating vertebrate neuromuscular nicotinic receptors and increasing cholinergenic transmission. As such it has potential as a treatment of cognitive function loss. A synthetic analog, DMXBA (GTS-21) has exhibited memory enhancing effects in recipients. The compound is currently under license by the Japanese pharmaceutical company Taiho and is in Phase I trials for treating Alzheimer's disease.

Microbe Derived Compounds

Cryptophycins

This family of compounds was initially reported from a terrestrial source (the *Nostoc* sp. cyanophyte component of a Scottish lichen). Representative compounds have also been found in species of free-living marine and non-marine cyanophytes and, more recently, from the Japanese sponge *Dysidea arenaria*. The expressed bioactivity that was originally pursued was activity as an antifungal agent. Development of the compound for this purpose was not pursued beyond preliminary investigations when it was deemed too toxic for human use. More recently it has been reported that cryptophycin 1 is an inhibitor of tubulin assembly in cells.

Curacin A

First isolated from *Lyngbya majuscula* by Gerwick et al. and reported in 1994, this natural product appeared to be a very potent tubulin interactive compound. It proved to be highly insoluble, however, so much so that bioactivity could not be demonstrated with *in vivo* animal models. These compounds are currently undergoing preclinical evaluation as potential future drugs.

Thiocoraline

Belonging to the chemical family known as the thiodepsipeptides, thiocoraline was first isolated from *Micromonospora marina*, an actinomycete bacterium collected from coastal Mozambique, southeast Africa. The Spanish marine drug company PharmaMar has reported that the compound shows activity against several standard drug screens, including breast cancer, colon cancer, renal cancer, and melanoma. Target cells appear to be inhibited through inhibition of DNA polymerase ýÿ enzyme. The most recent published literature suggests thiocoraline is still undergoing advanced preclinical evaluation.

The Future of Marine Bioprospecting

In the near future, marine bioprospecting efforts will likely focus not only on natural products from ocean plants, animals, and microbes, but also on the potential for biotech to exploit the information stored in the genomes of these organisms. In the wake of the Human Genome Project and with the expertise and technology that emerged from it, scientists are sequencing increasing numbers non-human genomes. Total genome sequencing of a handful of marine microbes has been completed, and the first sequencing of marine vertebrate (e.g., fish) genomes has commenced as well. This and similar research will likely lead to the development of gene probes that can identify the genes in target organisms that code for the elaboration of novel natural products of potential value to humans. Increasingly, new projects are expected to target the as yet unrealised biomolecular potential of the vast and almost entirely unknown marine microbial community. These research efforts could be instrumental in developing the next generation of pharmaceuticals for improving human health, as well as contributing to improved animal and agricultural crop health.

In summary, the marine world has become an important source of therapeutic agents with novel mechanisms of action. Even though thousands of new molecules are discovered every year only small number of candidates is incorporated in clinical trials. The main problem underlying this is sustainable supply of these compounds from natural sources. To battle this problem various strategies are developed, such as mariculture or aquaculture of source organisms, development of synthetic analogues of active compounds, fermentation of microorganisms producing the compound, etc. Another possible solution is the use of genetic engineering to transfer the genes encoding the synthetic enzymes that produce the desired compound to microorganisms that can be grown in huge quantities. Development of these products and services, as well as the fundamental research from which they must be derived will be enhanced by greater dependence on interdisciplinary sciences such as pharmacology, chemical ecology, molecular biology, genomics, metagenomics, computational and combinatorial chemistry and biology.

The field of marine natural products is passing its discovery phase and moving to the second phase where understanding relationships and processes is driving the research towards novel drugs from the sea. Marine plants, animals and microorganisms will be the basis of new products and services important to technology in the future. With rich biodiversity and vast marine resources along the Indian coast, in the form of estuaries, creeks, deep seas and continental shelf, the opportunities for research in the area of marine drug development are endless.

Probiotics and its Applications in Aquaculture

India is blessed with enormous aquatic resources. In fact in a country like India, providing sustainable livelihood opportunities and food security for ever increasing rural population is perhaps the greatest national challenge, where aquaculture plays a pivotal role in meeting this challenge. As per UNO projection on population over the next 50 years, India will exceed China as the most populous country in the world and taking this alarming situation, we should assume agriculture along with aquaculture and livestock as the most appropriate farm activities and as a tool for poverty reduction, food security and overall rural development. So blue revolution can be a solution to feed this ever increasing population. Although aquaculture in India has shown a rapid progress during past few years but some major problems are hindering the progress path and disease being one of them. Hence for successful aquaculture we must ensure quality feed, good environment and disease free seeds and juveniles.

With intensification of aquaculture involving high stocking densities coupled with external inputs of feed and fertilizer, sometimes of doubtful quality, the fish farmer is now face to face with diseases of various kinds from protozoan to helminth parasites to fungal, bacterial and viral diseases and syndromes. This has resulted in increased use of chemicals and antibiotics that are anyway not a welcome addition to the human food through this important food chain. India suffered the worst disaster in mid eighties due to epizootic ulcerative syndrome (EUS) which lead to heavy mortality of fishes in ponds, tanks, rivers and reservoirs.

The country again suffered a major setback again in brackish water aquaculture when the most valuable shrimp crops meant for export were lost in almost all maritime states particularly Andhra Pradesh. This was due to white spot syndrome and due to this a large number of farms were closed and farmers gave up shrimp farming which lead to decline in foreign exchange through shrimp farming. Fish and shellfish management have to be given their dues if the industry has to be sustainable and the farmers and entrepreneurs have to be making profits. Antibiotics have sometimes been used to reduce disease; however indiscriminate use has in some cases led to increased antibiotic resistance and problem of tissue residues and trade issues. There is a wide variety of vaccines available for other sectors but a very few in fisheries sector. The reason is very little research in fish pathology particularly in our country. So a question comes how to control disease? However, one measure that might be of assistance is the use of "probiotics". "Probiotics" generally includes bacteria, cyanobacteria, micro algae fungi, etc. Some Chinese researchers translate it into English as "Normal micro biota" or "Effective micro biota"; it includes Photosynthetic bacteria, *Lactobacillus, Actinomycetes, Nitrobacteria, Denitrifying* bacteria, Bifidobacterium, yeast, etc. Usually, it does not include micro algae. In English literature, probiotic bacteria are generally called the bacteria which can improve the water quality of aquaculture, and (or) inhibit the pathogens in water there by increasing production.

"Probiotics", "Probiont", "Probiotic bacteria" or "Beneficial bacteria" are the terms synonymously used for probiotic bacteria. Probiotics are a cultured product or live microbial feed supplement, which beneficially affects the host by improving its intestinal balance and health of the host. The first probiotic discovered long time ago was *Lactobacillus* sp., the lactic acid producing bacteria.

Current Research Status of Probiotics

Nogami and Maeda (1992) isolated a bacteria strain from a crustacean culture pond. The bacterial strain was found to improve the growth of crab (*Portunus trituberculatus*) larvae and repress the growth of other pathogenic bacteria, especially *Vibrio spp.,* but would not kill or inhibit useful micro algae in sea water when it was added into the culture water. Among the bacteria population present in the culture water of the crab larvae, the numbers of *Vibrio spp.* and pigment bacteria decreased or even became undetectable when the bacteria was added into culture water. The production and survival rate of crab larvae were greatly increased by the addition of the probiotic bacteria into the culture water. They also suggested that the bacterium might improve the physiological state of the crab larvae by serving as a nutrient source during its growth. This bacterium may have a good effect in the crab larval culture as a biocontrolling agent in the future.

Austin et al (1992) reported a kind of micro algae (*Tetraselmis suecica*), which can inhibit pathogenic bacteria of fish. *Teraselmis suecica* was observed to inhibit *Aeromonos hydrophila*, *A. salmonicida*, *Serrstia liquefaciens*, *Vibrio anguillaram*, *V. salmonicida* and *Yersnia ruckeri* type I. When used as a food supplement, the algal cells inhibited laboratory-induced infection in Atlantic salmon.

When used therapeutically, the algal cells and their extracts reduced mortalities caused by *A. salmonicida*, *A. liquefaciens*, *V. anguillaram*, *V. salmonicida* and *Yersnia ruckeri* type I. They suggested that there may be some bioactive compounds in the algal cells, and there appears to be a significant role for Tetraselmis in the control of fish diseases.

Smith and Davey (1993) reported that a fluorescent strain pseudomonad bacteria can competitively inhibit the growth of fish pathogen *A. salmonicida*.

Their results show that the fluorescent pseudomonad is capable of inhibiting the growth of *A. salmonicida* in culture media and that this inhibition is probably due to competition for free iron. In a challenge test of the Atlantic salmon by *A. salmonicida,* a statistically significant reduction in the frequency of stress-induced infection in the group of fish bathed in the bacterium fluorescent pseudomonad compared to the control group was observed.

Austin et al (1995) reported a probiotic strain of *Vibrio alginolyticus*, which did not cause any harmful effect in salmonids. By using the cross-streaking method, the probiont was observed to inhibit the fish pathogens. When the freeze-dried culture supernatant was added to the pathogenic bacteria such as *V. ordalii*, *V. anguillarum*, *A. salmonicida* and *Y. ruckeri*, showed a rapid or steady decline in the number of culturable cells, compared to the controls. Their results indicated that application of the probiont to Atlantic salmon culture led to a reduction in mortalities when challenged with *A. salmonicida* and to a lesser extent *V. anguillarum* and *V. ordalii*. The observation with this probiotic *Vibrio* is encouraging, and it appears that there is tremendous potential for the use of such probiotics in aquaculture as part of a disease control strategy Maeda and Nagami (1989) reported some aspects of the biocontrolling method in aquaculture. In their study bacterial strains possessing vibrio static activity which improved the growth of prawn and crab larvae were observed. By applying these bacteria in aquaculture, a biological equilibrium between competing beneficial and deleterious microorganisms was produced, and results show that the population of *Vibrio spp.*, which frequently causes large scale damage to the larval production, was decreased.

Survival rate of the crustacean larvae in these experiments showed much higher than those without the addition of bacterial strains. They hope that addition of these strains of bacteria will repress the growth of *Vibrio spp.*, fungi and other pathogenic microorganisms. Their data suggest that controlling the aquaculture ecosystem using bacteria and protozoa is quite possible and if this system is adopted, it will maintain the aquaculture environment in better condition, which will increase the production of fish and crustaceans.

Garriques and Arevalo (1995) reported that the use of *V. alginolyticus* as a probiotic agent may increase survival and growth in *P. vannamei* postlarvae by competitively excluding potential pathogenic bacteria, and can effectively reduce or eliminate the need for antibiotic prophylaxis in intensive larvae culture system.

They believe that in nature a very small percentage of *Vibrio sp.* is truly pathogenic, and the addition of potentially pathogenic bacteria to aquaculture system through water, algae, and/or Artemia was recognised. In their study, the addition of the bacteria *V. alginolyticus* as a probiotic to mass larvae culture tanks resulted in increased survival rates and growth over the controls and the antibiotic prophylaxes.

Jiravanichpaisal and Chuaychuwong *et al* (1997) reported the use of *Lactobacillus sp.* as the probiotic bacteria in the giant tiger shrimp (*P. monodon* Fabricius). They designed to investigate an effective treatment of *Lactobacillus sp.* against vibriosis and white spot diseases in *P. monodon*. They investigated the growth of some probiotic bacteria, and their survival in the 20 ppt sea water for at least 7 days. Inhibiting activity of two *Lactobacillus* sp. against *Vibrio sp.*, *E. coli*, *Staphylococcus sp.* and *Bacillus subtilis* was determined.

Direkbusarakom and Yoshimizu *et al* (1997) reported *Vibrio spp.* which dominate in shrimp hatchery against some fish pathogens. Two isolates of *Vibrio spp.* which are the dominant composition of the flora in shrimp hatchery, were studied for antiviral activity against infectious haematopoietic necrosis virus (IHNV) and Oncorhynchus masou virus (OMV). Both strains of bacteria showed the antiviral activities against IHNV and OMV by reducing the number of plaque. Their results demonstrate the possibility of using the *Vibrio* flora against the pathogenic viruses in shrimp culture.

Sugita and Shibuga (1996) reported the antibacterial abilities of intestinal bacteria in freshwater cultured fish. They isolated bacteria from the intestine of 7 kinds of freshwater cultured fish, and investigated the antibacterial abilities of these bacteria to 18 fish or human common pathogenic bacteria. Their results indicated that the bacteria isolated from intestine of 7 kinds of freshwater cultured fish possess the antibacterial abilities, and the presence of the intestinal bacteria can protect the fish against the infection by pathogenic bacteria.

Maeda and Liao (1992) reported on the effect of bacterial strains obtained from soil extracts on the growth of prawn larvae of *P. monodon*. Higher survival and molt rates of prawn larvae were observed in the experiment treated with soil extract, and the bacterial strain which promoted the growth of prawn larvae was isolated. They have assumed that if a specific bacterium is cultured and added to the prawn ecosystem to the level of 10 million cell/ml, other bacteria may hardly inhibit the same biotype because of protozoan activity which shall be one of the way to biologically control the aquaculture water biotype and ecosystem.

Maeda and Nogami *et al* (1992) have reported the utility of microbial food assemblages in culturing a crab, *Portunus trituberculatus*. Assemblages of microorganisms were produced by adding several nutrients, urea, glucose and potassium phosphate, to natural seawater with gentle aeration in which bacteria and yeast were prevailing. When these cultured microbes were added to sea water where crab larvae of *Portunus trituberculatus* were reared, bacteria numbers decreased very rapidly, followed by the decrease in flagellated protozoa and diatoms. Their results suggest that the crab larvae fed on these microorganisms successively. They found some strains of bacteria promoted larval growth, although yeasts did not support its growth. By adopting these assemblages of microorganisms a high yield was obtained for a prawn larva *P. japonicus*, although the success was not always consistent.

Douillet and Langdon (1994) have reported use of probiotics for the culture of larvae of the Pacific oyster (*Crassostrea gigas* Thunbeerg). They added probiotic bacteria as a food supplement to xenic larval cultures of the oyster *Crassostrea gigas* which consistently enhanced growth of larvae during different seasons of the year. Probiotic bacteria were added, at 0.1 million cells/ml, to cultures of algal-fed larvae, the proportion of larvae that are set to produce spat, and subsequently the number of spat increased. Manipulation of bacterial population present in bivalve larval cultures is a potentially useful strategy for the enhancement of oyster production. They suggest that the mechanisms of the action of probiotic bacteria are providing essential nutrients that are not present in the algal diets or improving the oyster's digestion by supplying digestive enzymes to the larvae or removing metabolic substances released by bivalves or algae.

Maeda and Liao (1994) have reported microbial processes in aquaculture environment and their importance in increasing crustacean production. They suggested that based on the photosynthesis of micro algae mainly, it was clarified that bacteria, protozoa and other microorganisms from microbial food assemblages use the organic matter produced by the algae and that these assemblages play a significant role in the aquatic food chain. The growth of the larvae and their production were markedly promoted by the probiotic bacteria. In their paper, they also described the presence of a bacterial clump, stained with a fluorescent dye, inside the digestive organ of the crab *Portunus trituberculatus*.

In China, the studies on probiotics in aquaculture were focused on the photosynthetic bacteria. Qiao Zhenguo et al (1992) have studied three strains of photosynthetic bacteria used in prawn (*P. chinensis*) diet preparation and their effect. Addition of the photosynthetic bacteria in the food or culture water was found to improve the growth of the prawn and the quality of the water. Cui Jingjin et al (1997) have reported on the application of photosynthetic bacteria in the hatchery rearing of *P. chinensis.* They used a mixture of several kinds of photosynthetic bacteria (*Rhodomonas sp.*) as water cleaner and auxiliary food.

Their results showed that the water quality of the pond treated with the bacteria was remarkably improved, the fouling on the shell of the larvae was reduced, the metamorphosis time of the larvae was 1 day or even earlier, and the production of post-larvae was more than that of the control.

Wang Xianghong *et al* (1997) have done some research work on probiotic bacteria in shrimp aquaculture. On the basis of studies on intestinal micro flora of wild adult shrimp *P. chinensis,* they have chosen some probiotic bacteria from shrimp intestinal flora. When the two probiotic bacterial strains were added to the larval culture water, the survival rate, the abilities of disease resistence and low salinity tolerance were improved; average body length and weight were increased. In addition, the probiotic bacteria, when added to the larval culture water was found not to influence the total bacterial number and water quality of the sea water. They also found that some probiotic bacteria can produce some digestive enzymes; these enzymes may improve the digestion of shrimp larvae, thus enhancing the ability of stress resistance and health of the larvae.

Mechanism of Action of the Probiotic Bacteria

The mechanism of action of the probiotic bacteria has not been studied systematically. According to some recent publications, in the aquaculture the mechanism of action of the probiotic bacteria may have several aspects.

1. Probiotic bacteria may competitively exclude the pathogenic bacteria or produce substances that inhibit the growth of the pathogenic bacteria.
2. Provide essential nutrients to enhance the nutrition of the cultured animals.
3. Provide digestive enzymes to enhance the digestion of the cultured animals.

4. Probiotic bacteria directly uptake or decompose the organic matter or toxic material in the water improving the quality of the water.

Chinese researchers have done some studies on the probiotic bacteria to improve the shrimp culture water, and achieved remarkable results (Li Zhuojia et al 1997). For example, when photosynthetic bacteria was added into the water, it could eliminate the NH3-N, H2S and organic acids, and other harmful materials rapidly, improve the water quality and balance the pH. The heterotrophic probiotic bacteria may have chemical actions such as oxidation, ammoniafication, nitrification, denitrification, sulphurication and nitrogen fixation. When these bacteria were added into the water, they could decompose the excreta of fish or prawns, remaining food materials, remains of the plankton and other organic materials to CO2, nitrate and phosphate.

These inorganic salts provide the nutrition for the growth of micro algae, while the bacteria grow rapidly and become the dominant group in the water, inhibiting the growth of the pathogenic microorganisms. The photosynthesis of the micro algae provide dissolved oxygen for oxidation and decomposition of the organic materials and for the respiration of the microbes and cultured animals. This kind of cycle may improve the nutrient cycle, and it can create a balance between bacteria and micro algae, and maintaining a good water quality environment for the cultured animals.

The Feasibility and Future of the Application of Probiotics in Aquaculture

Based on the previous research results on probiotics we suggest that the use of probiotic bacteria in aquaculture has tremendous scope and the study of the application of probiotics in aquaculture has a glorious future. At present, the probiotics are widely applied in United States of America, Japan, European countries, Indonesia, India and Thailand, with commendable results. The probiotics have become commodities in some countries, for example, the Alken-Murray Corporation and American Standard Products company of United States of America and the company of Japan have their probiotics products. The study may create a new field of industrial products, like the industrial fields of aquaculture product processing and Aquacultural food processing.

China is a large country in aquaculture, but the application and development of the probiotics in Chinese aquaculture is very meager when compared to other countries. In recent years, the diseases of shrimps hindered the development of shrimp culture. The Chinese government has realised the economic value and potential social benefits of the application of probiotics in aquaculture, and has, recently, paid more attention to the study and development of probiotics in aquaculture.

Thus the government has increased the research funds for it. Probiotics principally inhibit the growth and decrease the pathogenicity of the pathogenic bacteria, enhance the nutrition of the aquacultured animals, improve the quality of the aquaculture water and decrease the use of antibiotics and other chemicals; thus decreasing environmental contamination by the residual antibiotics and chemicals. This benefit of probiotics will be long lasting, and the application of probiotics will become a major field in the development of aquaculture in the future.

Application of Probiotics in Aquaculture

During the past 20 years, aquaculture industry has been growing tremendously, especially that of marine fish, shrimps and bivalves. But, as with many other industries, this rapid growth has brought with it the problem of environmental pollution. Contamination of coastal waters due to aquaculture is posing serious concerns among law makers as well as scientists. The coastal environment has been seriously damaged, often resulting in disease outbreaks.

Recently, shrimp culture all over the world has been frequently affected by viral and bacterial diseases inflicting huge loss. In China, the production of shrimps decreased seriously. The production of shrimps was 200,000 tons in 1992, but was only 55,000 tons in 1994. Pathogenic microorganisms implicated in these outbreaks were viruses, bacteria, rickettsia, mycoplasma, algae, fungi and protozoan parasites. For preventing and controlling diseases, a host of antibiotics, pesticides and other chemicals were used possibly creating antibiotic resistant bacteria, persistence of pesticides and other toxic chemicals in aquatic environment and creating human health hazards. Thus, how to improve the ecological environment of aquaculture has become the focus of attention of international aquaculture.

Now, researchers are trying to use probiotic bacteria in aquaculture to improve water quality by balancing bacterial population in water and reducing pathogenic bacterial load. Researchers are increasingly paying more attention to this new approach (ecological aquaculture), and have made considerable headway. This review, on the basis of the new research findings in probiotics applied to aquaculture, analyse and summarize the mechanism of probiotic action in aquaculture.

Use of Immunostimulants in Aquaculture Systems

The contribution of aquaculture to fish production is steadily increasing. The increase would have been much more but for the major constraint of loses in culture production, particularly of shrimp, due to diseases. In India, the loss of shrimp production during 1995-96 due to diseases is estimated to be Rs.600 crores and the loses continue around this level sine then. Loses in production of cultured shrimp have led to the realisation that the goal of aquaculture is not merely to increase production but to make it sustainable. In recent years the application of vaccination in respect to finfish and immunostimulants in respect of shrimp/finfish for disease management in aquaculture is being increasingly recognised. Generally, immunostimulants enhance individual components of the non specific immune response but this does not always translate into increased survival. I addition, immunostimulants fed at too high dose or for too long can be immunosuppressive.

The substances of capable of stimulating immune response are the compounds that promote release of from immune effecter cells. Immunostimulants enhance the humoral and cellular response in both specific and non-specific ways. These agents are widely used for impaired immune function and to stabilize the improved immune status. The use of immunostimulants in fish culture or in aquaculture of other species for prevention of diseases is a promising new development. In general, immunostimulants comprise a group of biological and synthetic compounds that enhance the non-specific defence mechanisms in animals, thereby imparting generalized protection. This protection may be particularly important for fish that are raised in or released into environments where the nature of pathogen is unknown and immunization by specific vaccine may be futile. Several immunostimulants have been evaluated in fin fishes.

Many occasions arise in the course of fish culture that calls for enhancement of immune response. These include strengthening of the normal immune response in order to enhance protection and reduce immunosuppressive conditions. Immunostimulants can be classified into several categories by their origin and mode of action—

1. bacteria and bacterial products,
2. complex carbohydrates,
3. vaccines,
4. immunity enhancing drugs,
5. nutritional factors,
6. animal extracts,
7. cytokines, and
8. Lectins, plant extracts.

Two main procedures for evaluating the efficiency of an immunostimulants are:

1. *In vivo*, eg., protection tests against fish pathogens: and
2. *In vitro,* eg, measurement of the efficiency of cellular and humoral immune mechanism.

Attention is focused on the lymphocyte proliferation test as an adequate method for providing a correct evaluation of the cellular immune condition which can be adopted together with the more commonly used parameters, such as phagocytosis and respiratory burst. It may be mentioned here that the use of immunostimulants in the diets of marine fish and the evaluation of their effect on the immune system of fish has been investigated. As immunostimulants, as such, which can be useful in preventing diseases in land based aquaculture, in pens and hatcheries rarely occurs alone in the natural environment, the subject deserves a discussion here.

Specific and Non-Specific Immunostimulants: Specific immunostimulation is related to the potentiation of the host's immune system towards a unique specific antigen. Vaccination is perhaps the best example of producing specific immunity. Non specific immunostimulation generally is an attempt to upgrade immunologic capabilities at a time when an animal may be exposed to one or several pathogens and/or be immuno-compromised.

Characteristics of an Ideal Immunostimulants

These can be described as:

1. It should be non-toxic, even at a high dose rate.
2. It should be non-carcinogenic or have long term side effects.
3. At therapeutic levels, it should have a short withdrawal period with low tissue residues.
4. It should stimulate a wide range of non-specific immune responses against bacteria, fungi, virus, protozoa and helminthes.
5. It should be capable of amplifying primary and secondary immune responses to infectious agents.
6. Breakdown products of compound concerned should be either inactive or readily biodegradable in the environment.
7. It should be having defined chemical composition or biological activity.
8. It should be active by oral route and should be stable both in its native state and after incorporation into food and water.
9. It should be compatible with arrange of drugs including antibiotics and anthelmintics, and
10. It should be inexpensive and either tasteless or palatable.

Objectives of Immunostimulation

These are:

1. Promoting a greater and more effective sustained immune response to those infectious agents producing subclinical disease without risks of toxicity, carcinogenicity or tissue residues.

2. Hastening the maturation of non-specific and specific immunity in young susceptible animals.
3. Enhancing the level of duration of specific immune response, both cell mediated and humoral, following vaccination.
4. Overcoming of immunosuppressive effects of stress and of those infectious agents that damage or interface with the functioning of cells of immune system.
5. Selectively stimulating the relevant components of the immune system or non-specific immune mechanism that preferentially confer protection against micro-organisms. For example via interferon release, especially for those infectious agents for which no vaccines currently exists; and
6. Maintaining immune surveillance at hightened level to ensure early recognition and elimination of neoplastic changes in tissues.

Some Common Immunostimulants

Muramyl Dipeptide: Muramyl dipeptide is a simple glycoprotein, also a purified form of mycobacteria. Its activity includes:

1. Enhancement of antibody activity.
2. Stimulation of polyclonal activation of lymphocytes, and
3. Activation of macrophages.

Levamisole: It is an anthelmintics chemical that has been shown to have some stimulating effect on the immunological reactivity of animals and humans. Activities of this agent are:

1. enhancement of cell mediated cytotoxicity, lymphokine production and suppressor cell function, and
2. Stimulation of pathagocytic activity of macrophages and neutrophils.

Glucans: Glucans are the most popular immunostimulants used in aquaculture. It is derived from yeast cell wall and from certain higher plants. It has excellent immunostimulatory properties and works well when injected or fed to fish.

Yano et al. (1991) showed that ýÿ3-1, 6, branched ýÿ3-1, 3 Glucans were effective in carp. Jenny and Anderson (1993) showed that the use of Glucans increased activity in non-specific defence mechanism and in protection against *Yesinia ruckeri*. Glucan treatment of Atlantic salmon (*salmo salar*) induced protection against *Vibrio salmonicidia*. Several Glucan products such as vitastim, macrogard, are marketed commercially and are used in supplementing fish feeds.

Chitin and Chitosan: Both chitin and chitosan have a major role in aquaculture. They are non-specific immunostimulators which are effective on a short term basis. Anderson & Swicki (1994) administered chitosan to brook trout (*Salvenus fontinalis*) by injection and immersion and found that high levels of protection occurred 1, 2, 3 days afterwards, but protection was greatly reduced by day 14. Injection of chitosan was also more effective than simple immersion.

Actually chitosan is a deacetylation product of chitin. The influence of chitosan on immune response of healthy and cortisol treated rohu was demonstrated. After treatment with chitosan sufficiently higher responses in almost all assays of non-specific immunity was observed in comparison to their healthy control or cortisol treated counterparts respectively without chitosan treatment (Sahoo and Mukherjee, 1999).

In aquaculture, chitosan has been used as an immunostimulant for protection against bacterial disease in fish, for controlled release of vaccines, and as a diet supplement (Bullock et al., 2000). Similar dose of chitosan in brook trout has been shown to be immunopotent. It had a higher degree of protection against *A. salmonicida* infection for a short duration. It also gave protection when feeding was done @ 0.5 gm/100gm feed for one week.

Vitamin C and E: both the vitamins are antioxidants. Vitamin C acts as a multiple cell stimulator. Diet supplemented with vitamin C gave protection against *A. salmonicida* in Atlantic salmon.

Vitamin E Stimulates B and T Lymphocytes: The mode of action of vitamin E in enhancing immunity is nuclear but it has been found that supplemental vitamin E may serve as a significant stimulus of immunity in some individuals.

Bacillus Calmette Guarine (BCG): It is a potent cytokine synthesis enhancer. It is actually a live attenuated vaccine strain of Mycobacterium bovis. BCG produces a generalized enhancement of both B cell and T cell mediated responses of phagocytosis and resistance to infection.

Streptococcal Components: These components are potent immunostimulants. Products from *Bordetella pertuosis, Brucella abortus, Bacillus subtilis* and *Klebsiella* pneumoniae all have immunostimulating activities.

Acemannan: It is a complex carbohydrate. It is a potent cytokine synthesis enhancer with anti-tumor and anti-viral activities. It also has the important property of stimulating wound healing.

Lentinan: It is a polysaccharide extracted from a comestible mushroom. *Lentinus elodes* is endowed with anti-tumor activity. Lentinan might act by increasing sensitivity to histamine and serotoxin.

Leaf Extract of Ocimum Sanctum: Effect of leaf extract Ocimum sanctum on:

1. the specific and non-specific immune responses and
2. Disease resistance against *Aeromonas hydrophila* was investigated in *Oreochromis mossambicus.*

It stimulated both antibody response and neutrophil activity. Dietary intake also enhances the antibody response and disease resistance to *Aeromonas*. Possibility of using *O. sanctum* as immunostimulant is used in the maintenance of finfish health in intensive freshwater aquaculture/

C-UP 111: It has immunostimulant activating leucocyte functions. Highest preventive effect has been shown against *Aeromonas* infection in Nile tilapia with improved neutrophil function, in comparison with Glucans and lactoferin. It is the most popular agent in aquaculture system.

Aquatim: This is a kind of immunostimulant developed by the department of Microbiology of the College of Fisheries, Mangalore and now manufactured and marketed by Mangalore Biotech Laboratory, Mangalore under this trade name. It is widely marketed in India, particularly in Karnataka and Goa.

Aerobic Coryneforms: *Propionebacterium acenes* promotes antibody formation when administered as a killed suspension. This bacteria is phagocytosed by macrophage and stimulates cytokine synthesis. This organism has a general immunostimulating action leading to enhanced antibacterial and antitumor activity.

Prophylactic Measures Used in Aquaculture

World aquaculture has grown tremendously during the last years becoming an economically important industry (Subasinghe *et al.*, 2009). Today it is the fastest growing food-producing sector in the world with the greatest potential to meet the growing demand for aquatic food (FAO 2006). Globally, aquaculture is expanding into new directions, intensifying and diversifying. A persistent goal of global aquaculture is to maximize the efficiency of production to optimize profitability.

However, disease is a primary constraint to the growth of many aquaculture species and is now responsible for severely impeding both economic and socio-economic development in many countries of the world. Disease caused by Vibrio spp. and Aeromonas spp. Are commonly episodes of mortality. When faced with disease problem the common response has been to turn to following prophylactic measures.

The following components are undertaken in prophylactic measurre for disease prevention in aquaculture ;

1. Prebiotics
2. Probiotics
3. Vaccine
4. Immunostimulants

Prebiotic

Prebiotics are a non digestible food ingredient that beneficially affects the host by selectively stimulating the growth and/or activity of one or a limited number of bacteria in the colon and thus improves host health. (Gibson and Roberfroid, 1995). In different studies since 1999, many substances have been investigated as prebiotic. Based on the study of Mahious and Ollevier (2005), Fooks *et al.* (1999), and Gibson *et al.* (2004), any foodstuff that reaches the colon, e.g. non-digestible carbohydrates, some peptides and proteins, as well as certain lipids, is a candidate prebiotic. Certain non-digestible carbohydrates seem authentic prebiotics. They include resistant inulin and oligofructose, transgalactooligosaccharides (TOS), lactulose, isomalto oligosaccharides (IMO), lactosucrose, xylo-oligosaccharides (XOS), soyabean oligosaccharides and glucooligosaccharides. From in vivo and in vitro studies, inulin and oligofructose, TOS and lactulose are presently classified as prebiotics. IMO, lactosucrose, XOS, soyabean oligosaccharides and glucooligosaccharides are not considered as functional ingredients since they do not fulfill all criteria for classification as prebiotics. Prebiotics are selectively fermented by Bifidobacteria, Lactobacillus and Bacteroides. Inclusion of prebiotic in the diet has been reported to increase the uptake of glucose (Breves et al., 2001) and bioavailability of trace elements (Bongers and van den Heuvel, 2003).

In later use of prebiotics, they have the binding capacity therefore increasing the absorption of mineral such as calcium, magnesium and iron; these minerals, are not absorbed in the small intestine and so reach the colon, where they are released from the carbohydrate matrix and absorbed.

Prebiotics have been reported to have numerous beneficial effects in fish such as increased disease resistance and improved nutrient availability. The reasons for the different results are not clear yet. It may be due to the different basal diet, inclusion level, type of omnosaccharide, adaptation period, chemical structure (degree of polymerization, linear or branched, type of linkages between monometric sugars), origin of prebiotic, animal characteristics (species, age, and stage of production), duration of use and hygienic conditions of the experiment. If beneficial effects of prebiotics are manifested in fishes, then prebiotics have much potential to increase the efficiency and sustainability of aquacultural production. Therefore, comprehensive research to more fully characterize the intestinal microbiota of prominent fish species and their responses to prebiotics is warranted and survival.

Probitics

The origin of the term probiotic is attributed to Parker (1974). The probiotics were defined as live microbial feed supplements that improve health of man and terrestrial livestock. The gastrointestinal microbiota of fish and shellfish are peculiarly dependent on the external environment, due to the water flow passing through the digestive tract. Most bacterial cells are transient in the gut, with continuous intrusion of microbes coming from water and food. Some commercial products are referred to as probiotics, though they were designed to treat the rearing medium, not to supplement the diet. This extension of the probiotic concept is pertinent when the administered microbes survive in the gastrointestinal tract. Otherwise, more general terms are suggested, like biocontrol when the treatment is antagonistic to pathogens, or bioremediation when water quality is improved. However, the first probiotics tested in fish were commercial preparations devised for land animals. Though some effects were observed with such preparations, the survival of these bacteria was uncertain in aquatic environment. Most attempts to propose probiotics have been undertaken by isolating and selecting strains from aquatic environment. These microbes were Vibrionaceae, pseudomonads, lactic acid bacteria, Bacillus spp. and yeasts. Three main characteristics have been searched in microbes as candidates to improve the health of their host. (1) The antagonism to pathogens was shown in vitro in most cases. (2) The colonization potential of some candidate probionts was also studied. (3) Challenge tests confirmed that some strains could increase the resistance to disease of their host. Many other beneficial effects may be expected from probiotics, e.g., competition with pathogens for nutrients or for adhesion sites, and stimulation of the immune system. The most promising prospects are sketched out, but considerable efforts of research will be necessary to develop the applications to aquaculture.

What are Aquatic Probiotics?

The concept for aquatic probiotics is a relatively new. When looking at probiotics intended for an aquatic usage it is important to consider certain influencing factors that are fundamentally different from terrestrial based probiotics. Aquatic animals have a much closer relationship with their external environment. There are the big differences between terrestrial and aquatic animals in the level of interaction between the intestinal microbiota and the surrounding environment. On the other hand, potential pathogens are able to maintain themselves in the external environment of the aquatic organisms and proliferate independently of the host (Hansen and Olafsen 1999; Verschuere et al. 2000; Kesarcodi-Watson et al. 2008). The bacterial community composition of the intestinal tract of

aquatic animals is different from that found in terrestrial animals, which the probiotic concept was developed. Man and terrestrial livestock undergo embryonic development within an amnion, whereas the larval forms of most fish and shellfish are released in the external environment at an early ontogenetic stage. These larvae are highly exposed to gastrointestinal microbiota-associated disorders, because they start feeding even though the digestive tract is not yet fully developed (Timmermans 1987), and though the immune system is still incomplete (Vadstein 1997).

Thus, probiotic treatments are particularly desirable during the larval stages (Gatesoupe 1999). The resident microbes benefit from a fairly constant habitat in the GI tract of man and terrestrial livestock, whereas most microbes are transient in aquatic animals (Moriarty 1990). These animals are poikilothermic and their associated microbiota may vary with temperature changes. Salinity changes in the rearing environment will also affect the microbiota and marine finfish are obliged to drink constantly to prevent water loss from the body. A consequence of the specificity of aquatic microbiota is that the most efficient probiotics for aquaculture may be different from those of terrestrial species (Gatesoupe 1999; Kesarcodi-Watson et al. 2008).

Defining probiotics is a challenge – even more so for aquaculture application. Historically, probiotics were defined according to their expected benefits or improvement to the host's intestinal balance. Being concerned with humans and terrestrial animals, probiotics were generally Gram-positive obligate or facultative anaerobes, mostly LAB. Based on the intricate relationship an aquatic organism has with the external environment when compared with that of terrestrial animals, the definition of probiotics for aquatic animals was modified at the end of the last century. Verschuere et al. (2000) defined aquatic probiotics as "Live microorganisms that have a beneficial effect on the host by modifying the microbial community, associated with the host, by ensuring improved use of the feed or enhancing its nutritional value, by enhancing the host response towards disease, or by improving the quality of its ambient environment". This implies a much wider range of microorganisms being used as probiotics for aquaculture animals that for terrestrial animals. The above definition is a more holistic and most appropriately defines probiotics for aquaculture.

Probiotics that currently used in aquaculture industry include a wide range of taxa – from Lactobacillus, Bifidobacterium, Pediococcus, Streptococcus and Carnobacterium spp. to Bacillus, Flavobacterium, Cytophaga, Pseudomonas, Alteromonas, Aeromonas, Enterococcus, Nitrosomonas, Nitrobacter, and Vibrio spp., yeast (Saccharomyces, Debaryomyces) and etc. (Irianto and Austin 2002; Burr et al. 2005; Sahu et al. 2008).

Aquatic Probiotics are Mainly of Two Types:

1) Gut probiotics which can be blended with feed and administrated orally to enhance the useful microbial flora of the gut .
2) Water probiotics which can proliferate in water medium and exclude the pathogenic bacteria by consuming all available nutrients. Thus, the pathogenic bacteria are eliminated through starvation (Nageswara and Babu 2006; Sahu et al. 2008).

The first type probiotics are using mainly in finfish aquaculture and the second type in shrimp aquaculture. Commercially available probiotics include pure strains, defined mixture of specific strains, but also consortia of strains and undefined mixtures. Generally, probiotics proposed as biological control agents in fish aquaculture are applied in the feed or as a water additive supplement.

Aquatic Probiotics are Marketed in Two Forms:

1) Dry forms: the dry probiotics that come in packets can be given with feed or applied to water. They have many benefits, such as safety, easy using, longer shelf life and etc. (Decamp and Moriarty 2007);

2) Liquid forms: the hatcheries generally use liquid forms which are live and ready to act. These liquid forms are directly added to hatchery tanks or blended with farm feed. The liquid forms can be applied any time of the day in indoor hatchery tanks, while it should be applied either in the morning or in the evening in outdoor tanks. Liquid forms give positive results in lesser time when compared to the dry and spore form bacteria, though they are lower in density (Nageswara and Babu 2006).

There are no reports of any harmful effect for probiotics but it is found that the biological oxygen demand level may temporarily be increased on its application; therefore it is advisable to provide subsurface aeration to expedite the establishment of probiotics organisms. A minimum dissolved oxygen level of 3% is recommended during probiotics treatment. The development of suitable probiotics for aquaculture is not a simple task. It requires empirical and fundamental research, full-scale trials as well as the development of appropriate monitoring tools and production under stringent quality control. A performing mixture of probiotic strains can be designed after evaluating the ability of individual strains to grow in low/high salinity under micro-aerophilic or anaerobic conditions, produce various enzymes, and more importantly, produce a range of inhibitory compounds (Decamp 2004).

Vaccine

A vaccine is any biologically based preparation intended to establish or to improve immunity to a particular disease or group of diseases. Vaccines have been used for many years in humans, terrestrial livestock, and companion animals against a variety of diseases.

Vaccines work by exposing the immune system of an animal to an "antigen"—a piece of a pathogen or the entire pathogen—and then allowing time for the immune system to develop a response and a "memory" to accelerate this response in later infections by the targeted disease-causing organism. Vaccines are normally administered to healthy animals prior to a disease outbreak. One analogy used for vaccines is that of an insurance policy (Komar et al 2004). A vaccine, if effective, can help prevent a future disaster from being a major economic drain. But vaccines, like insurance, have a premium, or cost. The producer must weigh the cost in materials and labour against the risk and cost of a disease outbreak to determine whether vaccination is warranted. When actual vaccine effectiveness is also unknown, this makes decision-making even more difficult. Consultation with a fish veterinarian or other fish health specialist will be helpful when examining cost vs. benefit of a particular vaccine.

The ideal vaccine:

a. is safe for the fish, the person(s) vaccinating the fish, and the consumer;
b. protects against a broad strain or pathogen type and gives 100% protection;
c. provides long-lasting protection, at least as long as the production cycle;
d. is easily applied;
e. is effective in a number of fish species;
f. is cost effective; and
g. is readily licensed and registered (Grisez and Tan 2005).

What are the Different Types of Vaccines?

There are many different types of vaccines, and new kinds are continuously under development. Of the types currently in use, the most common are described below. Bacterins are vaccines comprised of killed, formerly pathogenic bacteria. Bacterins stimulate the antibody-related portion of the immune response (i.e., the humoral immune response).

Live, attenuated vaccines are comprised of live micro-organisms (bacteria, viruses) that have been grown in culture and no longer have the properties that cause significant disease. Live attenuated vaccines will stimulate additional parts of the immune system (i.e., a cell-mediated, as well as a humoral [antibody] response). Toxoids are vaccines comprised of toxic compounds that have been inactivated, so they no longer cause disease. An example, used in humans, is the tetanus toxoid vaccine. Subunit vaccines are made from a small portion of a micro-organism (rather than the entire micro-organism) that ideally will stimulate an immune response to the entire organism.

Other types of vaccines in development use even more modern strategies. Examples include recombinant vector vaccines, which combine parts of disease-causing micro-organisms with those of weakened microorganisms, and DNA vaccines. Recombinant vector vaccines allow a weak pathogen to produce antigen. DNA vaccines are composed of a circular portion of genetic material that can, after being incorporated into the animal, produce a particular immune-stimulating portion of a pathogen (i.e., antigen) continuously, thus providing an "internal" source of vaccine material. Other vaccine strategies are also undergoing research and development.

How are Vaccines Given to Fish?

Vaccines are administered to fish in one of three ways: by mouth, by immersion, or by injection. Each has its advantages and disadvantages. The most effective method will depend upon the pathogen and its natural route of infection, the life stage of the fish, production techniques, and other logistical considerations. A specific route of administration or even multiple applications using different methods may be necessary for adequate protection.

Oral Vaccination results in direct delivery of antigen via the digestive system of the fish. It is the easiest method logistically because feeding is a normal, ongoing part of the production schedule. Stress on the fish is minimal, and no major changes in production are required.

Prior to feeding, vaccine is mixed, top-dressed, or bioencapsulated into the feed. To reduce leaching into the water and/or to provide some protection against breakdown of the vaccine by the fish's digestive processes, a coating agent is often used. For small fish (e.g., 1-5 g or less), bioencapsulation may be a preferred method of oral delivery. Live food (rotifers, brine shrimp) is added to a concentrated vaccine solution, and allowed to take up vaccine. This live food is then fed to fry or small fingerlings. Although oral vaccine is the most preferred method, it conveys relatively short immunity (compared to the other methods) such that additional vaccination may be required. In addition, because of the problems involved with getting the vaccine intact through the intestine and adequately stimulating the immune system, there are few commercial oral vaccines available (Komar et al. 2004).

Immersion Vaccination permits immune cells located in the fish's skin and gills to become directly exposed to antigens. These immune cells may then mount a response (e.g., antibody production), thus protecting the fish from future infection. Other types of immune cells in the skin and gills carry antigens internally, where a more systemic response will also develop. Immersion vaccination occurs by dip or by bath. Dips are short, typically 30 seconds, in a high concentration of vaccine. Baths are of longer duration—an hour or more—and in a much lower concentration of vaccine. In practice, dips are logistically more practical for large numbers of small (1- to 5-g) fish. Unfortunately, protection using immersion methods may not last long and a second vaccination may be required (Komar et al. 2004) because smaller, younger fish may have immature immune systems and because this is a more indirect route.

Injection Vaccination allows direct delivery of a small volume of antigen into the muscle (intramuscular (IM) injection) or into the body cavity (intracoelomic [ICe= intraperitoneal or IP] injection), allowing for more direct stimulation of a systemic immune response. Injection vaccines normally include an oil-based or water-based compound, known as an adjuvant, that serves to further stimulate the immune system. Injection is effective for many pathogens that cause systemic disease; and protection—6 months to a year—is much longer than by other methods. Every fish in the population is injected, giving more assurance to the producer. Another advantage is that multiple antigens (for different diseases) can be delivered at the same time. However, vaccination by injection is logistically the most demanding of all three methods. Fish must be anesthetized to minimize stress. Injection requires more time, labour, and skilled personnel. The correct needle size is important. The vaccine may incite a more severe reaction if it is injected into the wrong portion of the fish. And finally, smaller-sized fish (under 10 g) may not respond well to this method (Komar et al 2004).

Have Vaccines been Used in Fish?

Vaccines have been used in food fish, in particular the salmon industry, for approximately 30 years, and are believed to be one of the main reasons that salmon production has been so successful. Vaccination of salmon also dropped the industry's use of antibiotics to a mere fraction of its original use (Sommerset et al 2005). In Norway, for example, in 1987, before widespread use of vaccines, approximately 50,000 kg of antibiotics were used. By 1997, when vaccines had become more routine, antibiotic usage had dropped to less than 1000-2000 kg (Sommerset et al. 2005).

Diagnostic Tools Used in Fish Disease Diagnosis

Diagnosis is a label given for a medical condition or disease identified by its signs, symptoms, and from the results of various diagnostic procedures. The term "diagnostic criteria" designates the combination of signs, symptoms, and test results that allows the clinician to ascertain the diagnosis of the respective disease. It can be definition as "the recognition of a disease or condition by its outward signs and symptoms," or "the analysis of the underlying physiological/biochemical cause(s) of a disease or condition." Disease diagnosis is refers to the various procedures and techniques used to identify the nature of disease and to precisely pinpoint the primary and secondary pathogens involved. Proper diagnosis leads to accurate therapies and avoid indiscriminate use of chemotherapeutics.

Disease diagnosis is an integral part of aquatic health management. Proper diagnosis helps to adopt accurate therapy and avoid indiscriminate use of chemotherapeutics. Information on case history and clinical signs should be carefully used while examining samples for diagnosis.

Importance of Case History

Case history information is very vital for proper disease diagnosis and for taking proper remedial steps. During a disease outbreak, careful scrutiny of the case history information will help to precisely pinpoint the circumstances under which the disease has developed. Some of the case history that are taken in to account are,

- Water quality
- Feeding percentage
- Feed intake
- Fertilization schedule
- Time of last Algal bloom
- Liming details
- Treatment details
- Source of seed, stocking details

Classification of Disease Diagnostic Methods

The disease diagnostic methods can be classified in to following types

- Microscopic diagnosis
- Histological diagnosis
- Microbiological diagnosis
- Immunological diagnosis
- Molecular diagnosis

Microscopic Diagnosis

Without a microscope it is simply impossible to tell the difference between a water quality and a parasite problem. The microscope should be considered the most basic of tools in fish disease diagnosis. Microscopy can be done quickly, but accuracy depends on the experience of the microscopist and quality of equipment. Most specimens are treated with stains that colour pathogens, causing them to stand out from the background, although wet mounts of unstained samples can be used to detect fungi, parasites (including helminth eggs and larvae) and motile organisms. Visibility of fungi can be increased by applying 10% potassium hydroxide (KOH) to dissolve surrounding tissues and nonfungal organisms.

The stain is based on the likely pathogens, but no stain is 100% specific. Most samples are treated with Gram stain and, if mycobacteria are suspected, an acid-fast stain. However, some pathogens are not easily visible using these stains; if these pathogens are suspected, different stains or other identification methods are required because microscopic detection usually requires a microbe concentration of about 1 x 10/ml.

The microscopes that are commonly used in fish disease diagnosis are

1. Light microscope
2. Electron microscope

Light Microscope

The light microscope is the most commonly used microscope in disease diagnosis. In this microscope the scrapings from the fish can be directly visualized or a squash of organ can be prepared for the examination. It can also be used for interpreting histological slides and stained microorganisms.

Histological Diagnosis

Histology is a branch of biology that involves the microscopic examination of thin, stained tissue sections in order to study their structure and function and, in the case of histopathology, to determine changes which may be due to pathogens and disease. Histology and have a central role in disease diagnosis. Stained fish tissue sections are prepared and examined by light microscopy. Tissue changes resulting from infectious or non-infectious disease are identified and described. Immunohistochemical methods are also used to detect specific pathogens in tissue sections.

Preparation

Following sampling, fish tissues are placed in an aqueous fixative. This fixative preserves the morphology (structure and chemical constituents) of tissues and cells, so that they are capable of withstanding further preparatory steps without change. It is essential that tissues are fixed within a very short time after death to avoid disintegration of tissues or cells by the action of their own enzymes. Following fixation, tissues are gradually dehydrated to remove any tissue water, using a series of graded alcohols. The tissues are then 'cleared', which involves treatment with a substance that mixes completely with both the dehydrating fluid and the embedding agent. Next the tissues are embedded in molten paraffin wax and cooled to harden the wax so that thin sections can be cut using a microtome and then mounted onto glass microscope slides. The wax is removed from the sections before staining.

Staining

After clearing and rehydration, the tissue sections can be stained using biological stains or dyes. Haematoxylin and eosin (H&E) is the most widely used histological stain because of its ability to reveal a wide range of different tissue components. Gram's stain is a staining method for differentiating microorganisms. The technique is based on the capability of bacteria cell walls to retain the crystal violet dye in the Gram stain during solvent treatment.

The cell walls for Gram positive micro-organisms retain the primary violet as they have a higher peptidoglycan (sugars) and a lower lipid content than Gram negative bacteria. The Periodic Acid-Schiff (PAS) reaction is used to demonstrate certain carbohydrates that are present in some tissues, and provide identification of infecting fungus in fish tissues. The PAS positive sites stain pink/red.

Immunohistochemistry

Immunohistochemical staining methods have been developed for the detection of viruses such as infectious pancreatic necrosis virus (IPNV), infectious salmon anaemia virus (ISAV) and nodavirus in paraffin-embedded tissue sections. Viral antigen is localised by an antibody raised against the virus and subsequent detection steps result in a coloured product that can be visualised by light microscopy

Histological techniques enable the description of tissue pathology and highlight the sequence of cellular changes and their progression caused by infectious and noninfectious diseases. By examining stained sections, viruses, bacteria, fungi and parasites can be identified, and using immunohistochemical techniques, certain infectious agents can be detected in tissue sections. The increased use of image analysis tools by FRS allows qualitative data to be generated to enhance disease diagnosis.

Microbiological Diagnosis

Bacterial diseases can be identified by various microbiology methods. The samples are taken from the fish under aseptic condition and they are first grown in non-selective media, then the various microbial diagnostic methods are used. The different microbiological diagnostic methods are

1. Staining methods
2. Motility test
3. Culturing in selective medium
4. Biochemical test

Staining Method

Various staining methods are used to identify various bacterial pathogens among them gram staining is the most important and most commonly used method.

- *Gram stain:* The Gram stain classifies bacteria according to whether they retain crystal violet stain (gram-positive — blue) or not (gram-negative — red) and highlights cell morphology (eg, bacilli, cocci) and cell arrangement (eg, clumps, chains, diploids). Such characteristics can direct antibiotic therapy pending definitive identification. To do a Gram stain, technicians heat-fix specimen material to a slide and stain it by sequential exposure to Gram's crystal violet, iodine, decolorizer, and counterstain (typically safranin).
- *Acid-fast and moderate (modified) acid-fast stains:* These stains are used to identify acid-fast organisms (Mycobacterium sp) and moderately acid-fast organisms (primarily Nocardia sp). These stains are also useful for staining Rhodococcus and related genera, as well as oocysts of some parasites (eg, Cryptosporidium).
- *Fluorescent stains:* These stains allow detection at lower concentrations (1 10 cells/mL). Examples are acridine orange (bacteria and fungi), auramine-rhodamine and auramine O (mycobacteria), and calcofluor white (fungi, especially dermatophytes). Coupling a fluorescent dye to an antibody directed at a pathogen (direct or indirect immunofluorescence) should theoretically increase sensitivity and specificity. However, these tests are difficult to read and interpret, and few (eg, Pneumocystis and Legionella direct fluorescent antibody tests) are commercially available and commonly used.
- *India ink (colloidal carbon) stain:* This stain is used to detect mainly encapsulated fungi in a cell suspension.

The background field, rather than the organism itself, is stained, which makes any capsule around the organism visible as a halo. Leukocytes may appear encapsulated.

- *Wright's stain and Giemsa stain:* These stains are used for detection of parasites in blood, phagocytes and tissue cells, intracellular inclusions formed by viruses and some intracellular bacteria.
- *Trichrome stain (Gomori-Wheatley stain) and iron hematoxylin stain:* These stains are used to detect intestinal protozoa. The Gomori-Wheatley stain is used to detect microsporidia. It may miss helminth eggs and larvae. The iron hematoxylin stain differentially stains cells, cell inclusions, and nuclei. Helminth eggs may stain too dark to permit identification.

Motility Test

Histological techniques enable the description of tissue pathology and highlight the sequence of cellular changes and their progression caused by infectious and noninfectious diseases.

By examining stained sections, viruses, bacteria, fungi and parasites can be identified, and using immunohistochemical techniques, certain infectious agents can be detected in tissue sections.

The increased use of image analysis tools by FRS allows qualitative data to be generated to enhance disease diagnosis.

There are three methods to see the motility of the bacteria and they are

i. Hanging Drop Method
ii. Semi Solid Agar Method (Craigie)
iii. Three Coverslip Method

Hanging Drop Method

- Place a small drop of liquid bacterial culture in the centre of a coverslip
- Place a small drop of water at each corner of the coverslip
- Invert a slide with a central depression over the coverslip

- The coverslip will stick to the slide and when the slide is inverted the drop of bacterial culture will be suspended in the well
- Examine microscopically (x400) for motile organisms

Note: If well slides are not available, a ring of Vaseline or plasticine may instead be made on an ordinary microscope slide

Positive result: A darting, zigzag, tumbling or other organised movement

Negative result: No movement or Brownian motion only.

Semi Solid Agar Method (Craigie)

- Inoculate the test organism - the central glass tube
- Incubate at the relevant temperature for 18-24 hours
- Subculture from the outer section of the medium

Positive result: Organism can be recovered from the outer section of the medium Negative result: Organism remains in the inner tube

Culturing Bacteria Pathogen in Selective Medium

Certain bacteria have the capacity to grow in selective media. This property of bacteria is used to identify the bacterial pathogen. Only particular kind of bacteria can grow in selective media. Example for selective medium is TCBS for *Vibrios*.

Biochemical Tests

Various biochemical test are used for the identification of bacteria based on the character of a particular bacteria to give its positive and negative result for a particular biochemical test.

1. *The Indole Test:* The test organism is inoculated into tryptone broth, a rich source of the amino acid tryptophan. Indole positive bacteria such as Escherichia coli produce tryptophanase, an enzyme that cleaves tryptophan, producing indole and other products. When Kovaçs reagent (p-dimethylaminobenzaldehyde) is added to a broth with indole in it, a dark pink colour develops. The indole test must be read by 48 hours of incubation because the indole can be further degraded if prolonged incubation occurs. The acidic pH produced by Escherichia coli limits its growth.

2. *The Methyl Red and Voges-Proskauer Tests:* The methyl red (MR) and Voges-Proskauer (VP) tests are read from a single inoculated tube of MR-VP broth. After 24-48 hours of incubation the MR-VP broth is split into two tubes. One tube is used for the MR test; the other is used for the VP test. Media contains glucose and peptone. All enterics oxidize glucose for energy; however the end products vary depending on bacterial enzymes. Both the MR and VP tests are used to determine what end products result when the test organism degrades glucose. E. coli is one of the bacteria that produces acids, causing the pH to drop below 4.4. When the pH indicator methyl red is added to this acidic broth it will be cherry red (a positive MR test). Klebsiella and Enterobacter produce more neutral products from glucose (e.g. ethyl alcohol, acetyl methyl carbinol). In this neutral pH the growth of the bacteria is not inhibited. The bacteria thus begin to attack the peptone in the broth, causing the pH to rise above 6.2. At this pH, methyl red indicator is a yellow colour (a negative MR test). The reagents used for the VP test are Barritt's A (alpha-napthol) and Barritt's B (potassium hydroxide). When these reagents are added to a broth in which acetyl methyl carbinol is present, they turn a pink-burgundy colour (a positive VP test). This colour may take 20 to 30 minutes to develop.
3. *Catalase Test:* Catalase is the enzyme that breaks hydrogen peroxide (H_2O_2) into H_2O and O_2. Hydrogen peroxide is often used as a topical disinfectant in wounds, and the bubbling that is seen is due to the evolution of O2 gas. H_2O_2 is a potent oxidizing agent that can wreak havoc in a cell; because of this, any cell that uses O2 or can live in the presence of O_2 must have a way to get rid of the peroxide. One of those ways is to make catalase.
4. *The Citrate Test:* The citrate test utilises Simmon's citrate media to determine if a bacterium can grow utilising citrate as its sole carbon and energy source. Simmon's media contains bromthymol blue, a pH

indicator with a range of 6.0 to 7.6. Bromthymol blue is yellow at acidic pH's (around 6), and gradually changes to blue at more alkaline pH's (around 7.6). Uninoculated Simmon's citrate agar has a pH of 6.9, so it is an intermediate green colour. Growth of bacteria in the media leads to development of a Prussian blue colour (positive citrate). Enterobacter and Klebsiella are citrate positive while E.coli is negative.

5. *Oxidase Test:* The oxidase test identifies organisms that produce the enzyme cytochrome oxidase. Cytochrome oxidase participates in the electron transport chain by transferring electrons from a donor molecule to oxygen. The oxidase reagent contains a chromogenic reducing agent, which is a compound that changes colour when it becomes oxidized. If the test organism produces cytochrome oxidase, the oxidase reagent will turn blue or purple within 15 seconds.

6. *Nitrate Reduction Test:* Nitrate broth is used to determine the ability of an organism to reduce nitrate (NO_3) to nitrite (NO_2) using the enzyme nitrate reductase. It also tests the ability of organisms to perform nitrification on nitrate and nitrite to produce molecular nitrogen. Nitrate broth contains nutrients and potassium nitrate as a source of nitrate. After incubating the nitrate broth, add a dropperful of sulfanilic acid and a-naphthylamine. If the organism has reduced nitrate to nitrite, the nitrites in the medium will form nitrous acid. When sulfanilic acid is added, it will react with the nitrous acid to produce diazotized sulfanilic acid. This reacts with the a-naphthylamine to form a red-coloured compound. Therefore, if the medium turns red after the addition of the nitrate reagents, it is considered a positive result for nitrate reduction. If the medium does not turn red after the addition of the reagents, it can mean that the organism was unable to reduce the nitrate, or it could mean that the organism was able to denitrify the nitrate or nitrite to produce ammonia or molecular nitrogen. Therefore,

another step is needed in the test. If the medium does not turn red after the addition of the nitrate reagents, add a small amount of powdered zinc. Be careful, as powdered zinc is hazardous! If the tube turns red after the addition of the zinc, it means that unreduced nitrate was present. Therefore, a red colour on the second step is a negative result. The addition of the zinc reduced the nitrate to nitrite, and the nitrite in the medium formed nitrous acid, which reacted with sulfanilic acid. The diazotized sulfanilic acid that was thereby produced reacted with the a-naphthylamine to create the red complex. If the medium does not turn red after the addition of the zinc powder, then the result is called a positive complete. If no red colour forms, there was no nitrate to reduce. Since there was no nitrite present in the medium, either, that means that denitrification took place and ammonia or molecular nitrogen were formed.

7. *Urease Test:* Urease broth is a differential medium that tests the ability of an organism to produce an exoenzyme, called urease that hydrolyzes urea to ammonia and carbon dioxide. The broth contains two pH buffers, urea, a very small amount of nutrients for the bacteria, and the pH indicator phenol red. Phenol red turns yellow in an acidic environment and fuchsia in an alkaline environment. If the urea in the broth is degraded and ammonia is produced, an alkaline environment is created, and the media turns pink. Many enterics can hydrolyze urea; however, only a few can degrade urea rapidly. These are known as "rapid urease-positive" organisms. Members of the genus *Proteus* are included among these organisms. Urea broth is formulated to test for rapid urease-positive organisms. The restrictive amount of nutrients coupled with the use of pH buffers prevent all but rapid urease-positive organisms from producing enough ammonia to turn the phenol red pink.

8. *Phenol Red Broth:* Phenol Red Broth is a general-purpose differential test medium typically used to differentiate gram negative enteric bacteria. It contains peptone, phenol red (a pH indicator), a Durham tube, and one carbohydrate. We use three different kinds of phenol red broths. One contains glucose; one contains lactose, and the last contains sucrose. The objective of the exercise is to determine which organisms can utilise each sugar. Phenol red is a pH indicator which turns yellow below a pH of 6.8 and fuchsia above a pH of 7.4. If the organism is able to utilise the carbohydrate, an acid by-product is created, which turns the media yellow. If the organism is unable to utilise the carbohydrate but does use the peptone, the by-product is ammonia, which raises the pH of the media and turns it fuchsia. When the organism is able to use the carbohydrate, a gas by-product may be produced. If it is, an air bubble will be trapped inside the Durham tube. If the organism is unable to utilise the carbohydrate, gas will not be produced, and no air bubble will be formed.
9. *Casease Test:* Skim milk agar is a differential medium that tests the ability of an organism to produce an exoenzyme, called casease, that hydrolyzes casein. Casein forms an opaque suspension in milk that makes the milk appear white. Casease allows the organisms that produce it to break down casein into smaller polypeptides, peptides, and amino acids that can cross the cell membrane and be utilised by the organism. When casein is broken down into these component molecules, it is no longer white. If an organism can break down casein, a clear halo will appear around the areas where the organism has grown.
10. *Gelatinase Test:* Nutrient gelatin is a differential medium that tests the ability of an organism to produce an exoenzyme, called gelatinase, that hydrolyzes gelatin. Gelatin is commonly known as a component of gelled salads and some desserts, but it's actually a protein

derived from connective tissue. When gelatin is at a temperature below 32°C (or within a few degrees thereof), it is a semisolid material. At temperatures above 32°C, it is a viscous liquid. Gelatinase allows the organisms that produce it to break down gelatin into smaller polypeptides, peptides, and amino acids that can cross the cell membrane and be utilised by the organism. When gelatin is broken down, it can no longer solidify. If an organism can break down gelatin, the areas where the organism has grown will remain liquid even if the gelatin is refrigerated.

11. *Lipase Testl:* Tributyrin agar is a differential medium that tests the ability of an organism to produce an exoenzyme, called lipase, that hydrolyzes tributyrin oil. Lipases break down lipids (fats). Tributyrin oil is a type of lipid called a triglyceride. Other lipase tests use different fat sources such as corn oil, olive oil, peanut oil, egg yolk, and soybean oil. Lipase allows the organisms that produce it to break down lipids into smaller fragments. Triglycerides are composed of glycerol and three fatty acids. These get broken apart and may be converted into a variety of end-products that can be used by the cell in energy production or other processes. Tributyrin oil forms an opaque suspension in the agar. When an organism produces lipase and breaks down the tributyrin, a clear halo surrounds the areas where the lipase-producing organism has grown.

12. *Starch Hydrolysis:* Starch agar is a differential medium that tests the ability of an organism to produce certain exoenzymes, including a-amylase and oligo-1,6-glucosidase, that hydrolyze starch. Starch molecules are too large to enter the bacterial cell, so some bacteria secrete exoenzymes to degrade starch into subunits that can then be utilised by the organism. Starch agar is a simple nutritive medium with starch added. Since no colour change occurs in the medium when organisms hydrolyze starch, we add iodine to the

plate after incubation. Iodine turns blue, purple, or black (depending on the concentration of iodine) in the presence of starch. A clearing around the bacterial growth indicates that the organism has hydrolyzed starch.

13. *Triple Sugar Iron Agar:* Triple sugar iron agar (TSI) is a differential medium that contains lactose, sucrose, a small amount of glucose (dextrose), ferrous sulfate, and the pH indicator phenol red. It is used to differentiate enterics based on the ability to reduce sulfur and ferment carbohydrates. As with the phenol red fermentation broths, if an organism can ferment any of the three sugars present in the medium, the medium will turn yellow. If an organism can only ferment dextrose, the small amount of dextrose in the medium is used by the organism within the first ten hours of incubation. After that time, the reaction that produced acid reverts in the aerobic areas of the slant, and the medium in those areas turns red, indicating alkaline conditions. The anaerobic areas of the slant, such as the butt, will not revert to an alkaline state, and they will remain yellow. This happens with *Salmonella* and *Shigella*

14. *Decarboxylation Test:* Decarboxylase broth tests for the production of the enzyme decarboxylase, which removes the carboxyl group from an amino acid. Decarboxylase broth contains nutrients, dextrose (a fermentable carbohydrate), pyridoxal (an enzyme cofactor for decarboxylase), and the pH indicators bromcresol purple and cresol red. Bromcresol purple turns purple at an alkaline pH and turns yellow at an acidic pH. We also add a single amino acid to each batch of decarboxylase broth. The three amino acids we test in our decarboxylase media are arginine, lysine, and ornithine. The decarboxylase test is useful for differentiating the *Enterobacteriaceae*. Each decarboxylase enzyme produced by an organism is specific to the amino acid on which it acts. Therefore, we test the ability of organisms to produce arginine decarboxylase, lysine

decarboxylase, and ornithine decarboxylase using three different but very similar media. If the organism is unable to ferment dextrose, there will be no colour change in the medium. If an organism is able to ferment the dextrose, acidic byproducts are formed, and the media turns yellow. As the organisms ferment the dextrose, the media initially turns yellow, even when it has been inoculated with a decarboxylase-positive organism. The low pH and the presence of the amino acid will cause the organism to begin decarboxylation. If an organism is able to decarboxylate the amino acid present in the medium, alkaline byproducts are then produced. Arginine is hydrolyzed to ornithine and is then decarboxylated. Ornithine decarboxylation yields putrescine. Lysine decarboxylation results in cadaverine. These byproducts are sufficient to raise the pH of the media so that the broth turns purple. If the inoculated medium is yellow, or if there is no colour change, the organism is decarboxylase-negative for that amino acid. If the medium turns purple, the organism is decarboxylase-positive for that amino acid.

15. *Coagulase Test:* The coagulase test identifies whether an organism produces the exoenzyme coagulase, which causes the fibrin of blood plasma to clot. Organisms that produce catalase can form protective barriers of fibrin around themselves, making themselves highly resistant to phagocytosis, other immune responses, and some other antimicrobial agents. The coagulase slide test is used to identify the presence of bound coagulase or clumping factor, which is attached to the cell walls of the bacteria. Bound coagulase reacts with the fibrinogen in plasma, causing the fibrinogen to precipitate. This causes the cells to agglutinate, or clump together, which creates the "lumpy" look of a positive coagulase slide test. You may need to place the slide over a light box to observe the clumping of cells in the plasma. The coagulase tube

test has been set up as a demo for you to observe in class. This version of the coagulase test is used to identify the presence of either bound coagulase or free coagulase, which is an extracellular enzyme. Free coagulase reacts with a component of plasma called coagulase-reacting factor. The result is to cause the plasma to coagulate. In the demo, the coagulase plasma has been inoculated with *Staphylococcus aureus* and *Staphylococcus epidermidis* and allowed to incubate at 37°C for 24 hours. *Staphylococcus aureus* produces free coagulase; *Staphylococcus epidermidis* does not. The coagulase test is useful for differentiating potentially pathogenic *Staphylococci* such as *Staphylococcus aureus* from other Gram positive, catalase-positive cocci.

Bibliography

Adler-Nissen, J. : *Enzymic Hydrolysis of Food Proteins,* London: Elsevier Applied Science, 1985.

Allchin, F.R.: *The Agriculture of Early Historic South Asia,* Cambridge University Press, U. K., 1995.

Bagdi, Madan Lal : *Physiology, Biochemistry and Biotechnology,* Manglam Pub, Delhi, 2007.

Bjorndal, Trond: *The Economics of Salmon Aquaculture.*, Wiley-Blackwell, NY, 1990.

Bora, K.K. : *Agros Dictionary of Plant Physiology and Biochemistry,* Agrobios, Delhi, 2001.

Burgess, Peter: *Tropical Fishlopaedia: A Complete Guide to Tropical Fish Care,* Howell Books, NY, 2000.

Chopra, P.N.: *Agriculture of Ladakh: Our Cultural Fabric Series.* New Delhi: Ministry of Education and Social Welfare, 1978.

Collette B., and Facey D.: *The Diversity of Fishes,* Blackwell Publishing, UK, 2004.

Coultate, T. P. : *Food: The Chemistry of its Components,* London: The Royal Society of Chemistry, 1988.

Diamond, Jared: *How societies choose to fail or succeed,* Viking Press, 2005.

Doyle, Jack: *Agriculture, Genetics, and the Fate of the World's Food Supply,* Penguin Books, New York, NY, 1986.

Dubey, Pushpalata: *Fish Management and Aquatic Environment*, Daya, Delhi, 2006.

Ferguson, H.W.: *Systemic Pathology of Fish*, Iowa State Press, Ames, Iowa, 1989.

Francis, R.: *Galvanic Corrosion: A Practical Guide for Engineers,* NACE Press, 2001.

Grizzle J.M.: *Anatomy and Histology of the Channel Catfish*, Auburn Printing Co, 1976.

Heda, Nilesh: *Fresh Water Fishes of Central India: A Field Guide,* Vigyan Prasar, 2009.

Hepburn, J.: *Taking Aquaculture Seriously*, Organic Farming, Soil Association, 2002.

Hutchinson, Horace. G.: *Fly-Fishing in Salt and Freshwater*. London: John Van Voorst, 1851.

Kreeger, K.: *Down on the Fish Farm*: *Developing Effluent Standards for Aquaculture*, Bioscience, Uk, 2000.

Kulkarni, G.K.: *Fisheries and Fish Toxicology*, APH, Delhi, 2006.

LaQue, F.: *Marine Corrosion Causes and Prevention*, John Wiley and Sons, NY, 1975.

Lester R. Brown: *Stabilizing Climate*, W.W. Norton & Co., NY, 2006.

Lewbart G.A. : *Self-Assesment Color Review of Ornamental Fish*, Iowa State Press,1998.

Loiselle, P.V.: *The Cichlid Aquarium*, Tetra Press. UK, 1994.

Mori, Fumitashi: *Aquarium Fish of the World: The Comprehensive Guide to 650 Species*, Chronicle Books, UK, 1991.

Nash, Colin: *The History of Aquaculture,* John Wiley and Sons, NY, 2011.

Pollard, D. A. *Freshwater Fishes of South-eastern Australia.* McDowell, R. M. Reed, Sydney. 1980.

Poppe T.T., *A Color Atlas of Salmonid Diseases*, Academic Press, London, 1996.

Quinn, Thomas P.: *The Behavior and Ecology of Pacific Salmon and Trout,* University of Washington Press, Washington, 2005.

Rebelin W.E.: *The Pathology of Fishes*, The University of Wisconsin Press, UK, 1975.

Roth, Eva; Rosenthal, Harald; Burbridge, Peter: *Ecological Footprint' for Aquaculture Production,* Aquatic Living Resources, NJ, 2000.

Sandford, Gina: *Aquarium Owner's Guide*, DK Publishing, London, 2000.

Silva, De: *Fish Nutrition in Aquaculture*, Springer, Delhi, 2001.

Tietenberg TH: *Environmental and Natural Resource Economics: A Contemporary Approach*, Pearson/Addison Wesley, NY, 2006.

Turton, John: *The Angler's Manual; or Fly-fisher's Oracle*. London: R. Groombridge, 1836.

Wright, Leonard M. Jr.: *Fishing the Dry Fly As A Living Insect-An Unorthodox Method*. New York: E. P. Dutton & Co., 1972.

Index

A

Animal Welfare, 21.
Aquaculture, 19, 25, 34, 39, 42, 47, 50, 59, 62, 72, 75, 76, 123, 131, 142, 144, 260, 269, 276, 277, 283.
Aquaculture Biotechnology, 34.
Aquaculture Systems, 263.
Artificial Hybridization, 227, 300.

B

Bacterial Diseases, 97.
Bio-Remediation, 5.
Bioactive Natural Products, 240.
Biological Limitations, 127, 300.
Biotechnology, 1, 25, 26, 28, 32, 34, 35, 51, 61, 62, 63, 241, 298, 300.

C

Chromosome Engineering, 31.
Cnidarian Derived Compounds, 250.
Culturing Bacteria Pathogen, 284.

D

Derived Compounds, 247, 250, 251, 300.
Detecting Human Pathogens, 5.
Diagnostic Techniques, 70, 300.
Disease-Resistant Fish, 232.

F

Fish Disease Diagnosis, 278.
Fish Health Management, 32, 66.
Fish Production, 34, 41, 42, 45, 46, 47, 48, 49, 300.
Fisheries, 25, 41, 42, 43, 44, 72, 88, 239, 268.
Fisheries Sector, 41, 44.

G

Genetic Biotechnologies, 2, 4.
Genetic Variance, 164.
Genetically Modified fish, 6.

H

Histological Diagnosis, 280.
Human Health, 18.

I

Identifying Organisms, 64.
Improving Health, 61.
Individual Selection, 209.

L

Light Microscope, 280.
Living Marine Resources, 243.

M

Marine Ecology, 65.
Mechanism of Action, 260.
Microbiological Diagnosis, 281.
Microscopic Diagnosis, 279.

Mycotic Diseases, 105.

N

Natural Hybridization, 224.
Necropsy, 70, 300.
Nutritional Diseases, 79.

P

Phenotypic Variance, 163.
Potential, 8, 9, 57, 61, 243.
Probiotic Bacteria, 260.
Prophylactic Measures, 269.
Protozoa Infecting Gills, 81.

Q

Quantitative Phenotypes, 160, 185, 225, 300.

R

Regulation, Policy, 23.

S

Selective Breeding, 2, 187, 205, 240.
Selective Medium, 284.
Staining Method, 282.

T

Their Niches, 64.
Therapeutic Considerations, 72.
Transgenic Growth, 127, 300.

U

Understanding, 63, 144.

V

Vaccines, 5, 108, 274, 275, 276, 278.
Vertebrate-Derived, 247, 300.
Viral Diseases, 107.

W

Well-Being, 61.

Z

Zebrafish, 8, 14, 15, 16.

❑❑❑